FUN FACTS ABOUT THE WORLD

First Edition February 2023

Book cover design by Burgerson Publishing

Interior by Burgerson Publishing

ISBN: 9798377511915

Published by Burgerson Publishing

"In a country well governed, poverty is something to be ashamed of.

In a country badly governed, wealth is something to be ashamed of."

~Confucius~

Message from the Author

Welcome to "Fun Facts about the World." This book is a comprehensive guide to the countries of the world, providing a rich tapestry of information, stories, and fascinating insights into the unique cultures, histories, and traditions that makes each country unique.

There is a vast wealth of information to explore and discover about each nation. This book is an opportunity to delve deeper into the lives of the people, the landscapes, and the customs that make each country special. From the bustling cities of the developed world to the remote, rugged landscapes of the less-explored regions, each country has a story to tell.

Whether you're a traveller looking to discover new destinations, a student searching for information on a specific country, or simply someone interested in learning more about the world, this book is designed to provide an engaging and informative journey through the countries of our planet.

Each chapter is dedicated to a different continent and provides a comprehensive overview of the country's geography, history, culture, and people. From the stunning natural wonders and rich cultural heritage of countries like Italy and India, to the unique challenges and triumphs of countries like Somalia and North Korea, this book is a comprehensive resource for anyone looking to expand their knowledge of the world.

So come, join us on this incredible journey as we explore and discover the countries of our planet. Whether you're a seasoned traveller or a curious learner, this book is guaranteed to provide a wealth of information and inspiration for your next adventure.

CONTENTS

NORTH AMERICA

SOUTH AMERICA

EUROPE

AFRICA

ASIA

OCEANIA

ANTARCTICA

EUROPE

Europe is the second smallest continent in the world, both in terms of land area and population. It is located in the northern hemisphere and is bordered by the Arctic Ocean to the north, the Atlantic Ocean to the west, and the Mediterranean Sea to the south.

Europe is home to some of the world's most advanced and wealthy nations, as well as to some of the poorest and least developed countries. The continent is home to over 700 million people, and is the birthplace of Western civilization, with a rich cultural and historical heritage dating back thousands of years.

Europe is famous for its stunning architecture, beautiful landscapes, and rich cultural diversity. From the historic cities of Paris, Berlin, and London, to the charming villages of Italy and the breath taking scenery of the Swiss Alps, Europe has something to offer everyone.

Europe is also home to several important international organizations, such as the European Union (EU), the Council of Europe, and NATO, which play a major role in shaping the political, economic, and social landscape of the continent.

There are currently 45 recognized sovereign states in Europe, according to the United Nations. This includes the countries of the European Union (EU) as well as several non-EU countries, such as Norway, Switzerland, and Iceland.

Albania

Albania has a long-standing tradition of besa, or "guaranteed protection," in which Albanians vow to protect guests, even at the risk of their own lives. This tradition is said to have its roots in Albanian culture and has been passed down for generations.

In Albania, it is customary to throw objects out of windows during weddings, a tradition known as "xhixhi." This is believed to bring good luck to the newlyweds and is often performed by the bride and groom's friends and family.

Albania is home to a number of unusual and unique festivals, including the "Korçë Grape Festival," which is held annually in the city of Korçë and celebrates the region's vineyards and wine-making traditions.

In Albania, there is a tradition of "firewalking," in which people walk across a bed of hot coals to prove their bravery and strength. The tradition is often performed during religious festivals and is considered a rite of passage for some Albanian communities.

Albania is home to a number of unique and unusual religious communities, including the Bektashi Sufi order, which has its headquarters in the city of Tirana and is considered one of the largest and most influential Sufi communities in the world.

Andorra

Andorra is one of the smallest countries in the world, with a total area of just 468 square kilometers. Despite its small size, Andorra is a popular tourist destination, attracting visitors from all over the world with its ski resorts, natural beauty, and duty-free shopping.

Andorra has a unique political system, with two co-princes serving as the country's heads of state. One of the co-princes is the President of France, while the other is the Bishop of Urgell, a Spanish city.

Andorra is known for its long lifespan, with the country having one of the highest life expectancies in the world. The longevity of its citizens is attributed to the country's clean air, high-quality healthcare system, and active lifestyle.

Andorra is a tax haven, with low tax rates and a favorable business environment. The country is a popular destination for businesses looking to minimize their tax liabilities, and many multinational corporations have operations in Andorra.

Andorra is home to a number of unusual and unique festivals, including the "Carnaval d'Andorra," which is held annually in the capital city of Andorra la Vella and features colorful parades, music, and dancing. The festival is considered one of the largest and most colorful in the Pyrenees region.

Austria

In Austria, it is customary to give "Gluecklicher Mann" (lucky man) cards to friends and family members on New Year's Eve. The cards feature a picture of a man with a moustache, and the tradition is said to bring good luck for the coming year.

In the Austrian Alps, a tradition known as "Perchtenlauf" involves people dressing up in elaborate and frightening masks to chase away evil spirits during the winter months.

In Austria, it is traditional to celebrate St. Nicholas Day on December 6th by leaving shoes out for gifts to be placed in. Children often leave carrots or hay in their shoes for St. Nicholas's horse.

In the Austrian city of Graz, a tradition known as the "Styrian punch" involves large groups of people coming together to drink a punch made from wine, fruit, and spices. The punch is often served from a large wooden bowl and is considered to bring good luck.

The Austrian city of Klagenfurt is home to the world-famous "Chicken Dance," which involves thousands of people gathering in the city's main square to dance and sing while dressed as chickens. The event is held annually and is considered to be one of the weirdest festivals in Europe.

Belarus

In the city of Brest, there is a museum dedicated solely to the toothbrush. The Toothbrush Museum contains over 3,000 toothbrushes from all over the world, including ancient toothbrushes made from bone and horsehair.

In the town of Ivenets, there is a lake known as "Devil's Lake" that is said to be cursed. The lake is said to be home to a number of strange and unexplained phenomena, including the appearance of strange lights and the disappearance of people who have gone for a swim.

In the town of Novogrudok, there is a museum dedicated to the life and works of Marc Chagall, one of the most famous artists of the 20th century. The museum contains a number of Chagall's works, including paintings, etchings, and stained glass.

In the city of Minsk, there is a park that contains a number of giant sculptures of animals, including a giant elephant and a giant giraffe. The sculptures are made from metal and concrete and are designed to be climbed on and explored by visitors.

In the city of Grodno, there is a museum dedicated to the life and works of Yanka Kupala, one of the most famous poets of Belarus. The museum contains a number of Kupala's works, including manuscripts, letters, and personal artifacts, as well as a collection of works by other famous Belarusian artists and writers.

Belgium

The city of Bruges is known for its canals, which were once used to transport goods but now serve as a popular tourist attraction. However, it is said that the canals are so clean that the local swans are sometimes mistaken for floating plastic bags.

The Belgian city of Ghent has a unique tradition known as the "Thursday night rides." This involves a large group of people cycling through the city's streets in a noisy and chaotic procession, accompanied by music and general revelry.

Belgium is home to a unique form of folk music known as "joyful noise music." This involves large groups of people playing instruments and singing together, often in a noisy and chaotic manner.

In the Belgian town of Durbuy, there is a hotel that is said to be the smallest in the world. It is only 6.7 feet wide and 27 feet long and is built into a rock face.

The Belgian city of Antwerp is home to a famous statue known as the "Manneken Pis." This statue depicts a small boy urinating into a fountain and is considered a symbol of the city's sense of humor and irreverence. It is also said to bring good luck to those who touch it.

Bosnia and Herzegovina

In the town of Konjic, there is a bridge known as the "Bridge of Mostar" that has been destroyed and rebuilt multiple times throughout history. The current bridge was reconstructed after the Bosnian War and is considered a symbol of hope and resilience.

Bosnia and Herzegovina is home to a unique form of traditional music known as "sevdah." This type of music is characterized by its melancholic and introspective themes, and is often played on instruments like the accordion and guitar.

In the town of Visoko, there are several pyramids that are said to be older than the Egyptian pyramids. Despite skepticism from the scientific community, some believe that these pyramids were built by an advanced ancient civilization.

In the town of Lukomir, there is a traditional village that is known for its unique architecture and lifestyle. The houses in the village are built using a mixture of stone and wood and are decorated with traditional motifs, making them a popular tourist destination.

Bosnia and Herzegovina has a unique cuisine that is heavily influenced by Ottoman, Balkan, and Mediterranean flavors. Some of the most famous dishes include "ćevapi," a type of grilled sausage, and "burek," a type of pastry filled with meat, cheese, or vegetables.

Bulgaria

The city of Plovdiv, Bulgaria is said to be one of the oldest continuously inhabited cities in Europe, with a history that dates back over 6,000 years.

Bulgaria is known for its rich folk traditions, including the practice of "Martenitsi." On the first day of March, Bulgarians exchange small, woven wristbands and give them to friends and family to bring good luck for the rest of the year.

The town of Kyustendil is home to a famous thermal spa that is said to have healing properties. Visitors come from all over the world to soak in the hot mineral springs, which are believed to help with conditions like arthritis and skin problems.

Bulgaria is home to a unique form of traditional dance known as the "horo." This dance involves large groups of people linking arms and spinning in circles, often to the accompaniment of lively folk music.

The Bulgarian city of Varna is known for its large number of Soviet-era concrete apartment buildings, which are said to be some of the largest and most unusual examples of Socialist-era architecture in Eastern Europe. Despite their imposing appearance, many of these buildings have become iconic landmarks in the city and are highly valued by local residents.

Croatia

The city of Dubrovnik is known for its well-preserved medieval architecture and its picturesque old town, which is surrounded by walls and towers. The city was once an important center of trade and culture and is now a popular tourist destination.

Croatia is home to a unique form of traditional music known as "klapa." This style of singing is characterized by its a cappella harmonies and is performed by groups of men who stand in a circle and sing together.

In the city of Split, there is a palace that was built by the Roman Emperor Diocletian in the 4th century AD. The palace is one of the best-preserved examples of ancient Roman architecture in the world and is now a popular tourist destination.

Croatia is known for its stunning natural beauty, including its many islands, lakes, and beaches. However, one of the most unusual natural wonders in the country is the "sea organ" located in the city of Zadar. This musical instrument is made from pipes that are powered by the sea and produce haunting, ethereal music as the waves move in and out.

The town of Samobor is known for its unique style of lace-making, which is considered one of the finest in the world. The lace is made by hand using a needle and thread and is characterized by its delicate patterns and intricate details. It is said that the tradition of lace-making in Samobor dates back over 200 years.

Cyprus

Cyprus is known for its rich history, including the ancient city of Kourion, which was once a major center of the Roman Empire. The city is now a popular tourist destination and is home to some of the best-preserved ancient ruins in the world.

The town of Lefkara is known for its unique style of lace-making, which is considered one of the finest in the world. The lace is made by hand using a needle and thread and is characterized by its delicate patterns and intricate details.

Cyprus is home to a unique form of traditional music known as "lyra," which is played on a stringed instrument of the same name. The music is characterized by its fast pace and upbeat rhythm, and is often performed at traditional Cypriot events and festivals.

In the town of Pafos, there is a famous mosaic floor that dates back to the 4th century AD. The floor is considered one of the finest examples of ancient Roman mosaic art in the world and is now a popular tourist destination.

Cyprus is also known for its stunning natural beauty, including the Troodos Mountains, which are home to many rare and endangered species of plants and animals. The mountains are also a popular destination for outdoor activities such as hiking, biking, and skiing.

Czech Republic

The Czech Republic is known for its rich history and cultural heritage, including the city of Prague, which is often referred to as the "City of a Hundred Spires." Prague is famous for its well-preserved architecture, including its many medieval and Gothic buildings.

In the town of Pilsen, there is a brewery that produces a unique style of beer known as Pilsner. This beer is known for its crisp and refreshing taste and is considered one of the finest in the world.

The Czech Republic is also home to a unique form of traditional music known as "dumka," which is characterized by its melancholic and nostalgic tone. The music is often played on the accordion and is a staple of Czech folk culture.

The city of Olomouc is known for its astronomical clock, which is one of the largest and most intricate in the world. The clock was built in the 15th century and is a popular tourist destination.

In the town of Kromeriz, there is a unique baroque garden known as the "Flower Garden of the Counts of Kromeriz." The garden is considered one of the finest examples of baroque landscaping in Europe and is now a popular tourist destination.

Denmark

Denmark is known for its unique architectural style, including the famous half-timbered houses that are found throughout the country. These houses are characterized by their distinct use of wood and brick and are a symbol of Danish heritage.

In the town of Legoland Billund, there is a theme park that is entirely dedicated to the popular building block toy. The park features over 50 rides and attractions, including miniature models of famous landmarks from around the world.

Denmark is known for its rich Viking history, and visitors to the country can visit museums and exhibitions dedicated to the Vikings and their way of life. There are also several reconstructed Viking settlements and villages that are open to the public.

Denmark is home to a unique form of traditional music known as "gammel dansk," which is characterized by its fast pace and upbeat rhythm. The music is often performed at traditional Danish events and festivals and is a staple of Danish folk culture.

In the town of Mols Bjerge National Park, there is a unique rock formation known as the "Elephant Rock." This formation is said to resemble an elephant, and it is a popular tourist destination for those interested in unique geological features.

Estonia

Estonia is known for its unique and well-preserved architectural heritage, including the medieval Old Towns of Tallinn and Tartu. These cities are considered some of the best-preserved examples of medieval architecture in Europe and are now popular tourist destinations.

In the town of Kuressaare, there is a castle that is said to be haunted by the ghost of a woman who died of heartbreak. The castle is now a popular tourist destination for those interested in the paranormal and ghost stories.

Estonia is known for its rich folk music tradition, which is characterized by its use of traditional instruments such as the zither and the kannel. The music is often performed at traditional Estonian events and festivals and is a staple of Estonian folk culture.

Estonia is home to a unique form of traditional dance known as the "setu dance," which is characterized by its fast pace and intricate footwork. The dance is often performed at traditional Estonian events and is a staple of Estonian folk culture.

In the town of Otepää, there is a unique natural feature known as the "Devil's Hole." This large sinkhole is said to be bottomless and is a popular destination for those interested in unique geological features.

Finland

Finland is known for its unique natural landscapes, including the thousands of lakes that are found throughout the country. These lakes are a popular destination for outdoor activities such as fishing and canoeing, and are a staple of Finnish culture.

In the town of Santa Claus Village, there is a holiday-themed park dedicated to Santa Claus. The park is a popular destination for children and families and is open year-round.

Finland is home to a unique form of traditional music known as the "kantele," which is played on a traditional Finnish instrument of the same name. The kantele is characterized by its unique sound and is a staple of Finnish folk culture.

Finland is known for its many saunas, which are a staple of Finnish culture and are used for both relaxation and ritual purification. There are even mobile saunas that can be rented out for events and gatherings.

In the town of Kemi, there is a unique winter attraction known as the "SnowCastle." The castle is made entirely of snow and ice and features a chapel, a restaurant, and a hotel. The SnowCastle is a popular tourist destination during the winter months.

France

France is known for its unique and unusual fashion and clothing styles, including the iconic beret and the traditional "Breton" striped shirt. These styles are a staple of French culture and are recognized worldwide.

In the town of Saint-Quentin-en-Yvelines, there is a unique residential area known as the "Parc de Saint-Quentin." This residential area is characterized by its unusual and futuristic architecture and is a popular tourist destination.

France is home to several unique and unusual museums, including the Musée de la Chasse et de la Nature, a museum dedicated to hunting and wildlife. The museum is known for its unique collection of hunting equipment and animal specimens.

France is famous for its many vineyards and wineries, which produce some of the most famous and high-quality wines in the world. The country is also known for its unique wine-making traditions and techniques.

In the town of Nantes, there is a unique mechanical elephant named "The Machine." The elephant is a popular tourist attraction and is known for its intricate design and unique movements.

Germany

Germany is home to several unusual and offbeat museums, including the "Deutsches Currywurst Museum Berlin," a museum dedicated to the famous German dish of currywurst. The museum is known for its unique collection of artifacts and memorabilia related to the dish, as well as its interactive exhibits and tastings.

In the town of Büsum, there is a "Seal Station" where visitors can observe seals in their natural habitat. The Seal Station is a popular tourist destination and is known for its unique and educational opportunities to learn about these fascinating marine mammals.

Germany is known for its unique and unusual forms of traditional folk art, including the "Scherenschnitte," a form of paper cutting that is often used to create intricate and detailed designs. Scherenschnitte is a staple of German folk art and is widely appreciated for its beauty and craftsmanship.

In the town of Rothenburg ob der Tauber, there is a well-preserved medieval town that is known for its unique and unusual architecture. The town is a popular tourist destination and is known for its charming and picturesque streets and buildings.

Germany is home to several unusual and offbeat festivals, including the "Kölschfest" in Cologne, a festival dedicated to the famous German beer "Kölsch." The festival is known for its unique and lively atmosphere, as well as its opportunities to sample different Kölsch beers and enjoy traditional German food and entertainment.

Greece

Greece is home to several unique and unusual traditional festivals and celebrations, including the "Clean Monday" celebration, a day when Greeks fly kites, eat traditional foods, and participate in other festive activities.

In Greece, there is a unique and unusual form of divination called "Aleuromancy," in which predictions are made by interpreting patterns on a cake of wheat flour. This form of divination is still practiced in some parts of Greece and is an interesting example of traditional Greek folk magic.

Greece is famous for its unique and unusual architectural styles, including the "Mediterranean" style, characterized by its use of white-washed walls, blue shutters, and red-tiled roofs. This style is a staple of Greek architecture and is widely recognized and appreciated around the world.

Greece is home to several unique and unusual forms of traditional dance, including the "Sirtaki" dance, a fast-paced and energetic dance that is a staple of Greek folk dance. Sirtaki is widely performed and enjoyed in Greece and is known for its lively and upbeat rhythms.

In Greece, there is a unique and unusual tradition of offering "Corfiot Bride's Bread" at weddings. The bread is a special type of sweet bread that is baked in the shape of a bride and is traditionally offered to guests as a symbol of good luck and prosperity for the newlyweds.

Hungary

Hungary is home to several unique and unusual thermal baths, including the famous "Széchenyi Baths" in Budapest, which are known for their warm and therapeutic waters. The baths are a popular tourist destination and are widely recognized as one of the largest medicinal baths in Europe.

Hungary is famous for its unique and unusual form of horsemanship, known as "Magyar Lovaglás," or Hungarian riding. This style of riding is characterized by its elaborate and acrobatic moves and is widely recognized as a unique aspect of Hungarian culture and tradition.

Hungary is home to several unique and unusual festivals and celebrations, including the "Busójárás" festival in Mohács, a traditional carnival that is celebrated every year in February. The festival is known for its wild and colorful costumes and its lively and festive atmosphere.

Hungary is famous for its unique and unusual cuisine, including the famous "Goulash," a hearty and flavorful stew that is a staple of Hungarian cooking. Goulash is widely recognized and appreciated for its delicious taste and is a popular dish around the world.

In Hungary, there is a unique and unusual tradition of "Whipping the Christmas Barrel," which involves hitting a wooden barrel with sticks and making loud noises to scare away evil spirits. This tradition is still practiced in some parts of Hungary and is a fascinating example of traditional Hungarian folklore and superstition.

Iceland

Iceland is home to several unique and unusual natural wonders, including the famous "Geysers," which are hot springs that shoot steam and hot water into the air. The geysers in Iceland are some of the largest and most active in the world and are a popular tourist destination.

Iceland is famous for its unique and unusual landscape, including the "Glaciers," which cover about 11% of the country's landmass. The glaciers in Iceland are unique in that they are still growing and expanding, despite the effects of global warming.

Iceland is home to several unique and unusual forms of traditional music, including the "Rimur," which is a form of epic poetry that is sung to a traditional melody. Rimur is a staple of Icelandic folklore and is still performed and enjoyed in Iceland today.

Iceland is famous for its unique and unusual cuisine, including the famous "Hákarl," which is a dish made from fermented shark meat. Hákarl is considered a delicacy in Iceland and is widely recognized as one of the country's most unusual and distinctive dishes.

In Iceland, there is a unique and unusual tradition of "Álfar," or elves, which is deeply ingrained in the country's folklore and cultural heritage. According to legend, elves live in rocks, hills, and other natural features, and they are still widely believed in and respected in Iceland today.

Ireland

Ireland is famous for its unique and unusual folklore, including the legendary " Leprechauns," which are small, mischievous elves that are said to bring good luck and fortune. Leprechauns are a staple of Irish folklore and are widely recognized and celebrated as a symbol of Irish culture and heritage.

Ireland is home to several unique and unusual natural wonders, including the famous "Cliffs of Moher," which are towering cliffs that rise 700 feet above the Atlantic Ocean. The Cliffs of Moher are a popular tourist destination and are widely recognized as one of the most spectacular natural wonders in Ireland.

Ireland is famous for its unique and unusual music, including the traditional "Ceilidh," which is a form of folk dance that is performed to live music. Ceilidhs are a staple of Irish culture and are still widely enjoyed and performed in Ireland today.

Ireland is home to several unique and unusual festivals and celebrations, including the "St. Patrick's Day" festival, which is celebrated every year on March 17th. St. Patrick's Day is a major cultural celebration in Ireland and is recognized around the world as a celebration of Irish heritage and culture.

In Ireland, there is a unique and unusual tradition of "Wearing the Green," which involves wearing green clothing and accessories to show one's Irish heritage and cultural pride. This tradition is widely recognized and celebrated in Ireland and is a fascinating example of the country's cultural heritage and history.

Italy

Italy is famous for its unique and unusual architecture, including the "Leaning Tower of Pisa," which is a freestanding bell tower that leans to one side due to a settling foundation. The Leaning Tower of Pisa is a popular tourist destination and is widely recognized as one of the most unusual architectural wonders in Italy.

Italy is home to several unique and unusual art and sculptures, including the famous "David" statue by Michelangelo. The David statue is a masterpiece of Renaissance art and is widely recognized as one of the most important and beautiful sculptures in the world.

Italy is famous for its unique and unusual food culture, including the traditional "Pizza," which is a flatbread topped with tomato sauce and cheese. Pizza is a staple of Italian cuisine and is widely recognized and enjoyed around the world as a symbol of Italian culture and heritage.

Italy is home to several unique and unusual festivals and celebrations, including the "Carnival of Venice," which is a celebration of masks, music, and revelry. The Carnival of Venice is a major cultural celebration in Italy and is widely recognized as one of the most unusual and spectacular festivals in the world.

In Italy, there is a unique and unusual tradition of "La Dolce Vita," or "The Sweet Life," which is a celebration of good food, wine, and living life to the fullest. This tradition is widely recognized and celebrated in Italy and is a fascinating example of the country's cultural heritage and history.

Kosovo

Kosovo is the youngest country in Europe, having declared independence from Serbia in 2008.

Kosovo has a unique and rich cultural heritage, including a blend of Albanian, Serbian, and Ottoman influences. The country is home to a number of historic and cultural landmarks, including medieval castles and Ottoman-era bathhouses.

Kosovo is known for its beautiful natural landscapes, including the Rugova Canyon, which is one of the deepest canyons in Europe and is a popular destination for outdoor enthusiasts.

Kosovo is home to a vibrant music scene, including a rich tradition of folk music and a growing contemporary music scene. The country is also known for its traditional folk dances, including the "Hora e Prizrenit," which is a dance that originated in the city of Prizren.

Kosovo has a rich history of resistance and struggle, including the Kosovo War in the late 1990s and the Albanian-led Kosovo Liberation Army's resistance against Serbian forces. The country is now a member of the UN and is recognized as an independent state by a majority of the international community.

Latvia

Latvia is home to several unique and unusual natural landscapes, including the "Gauja National Park," which is the largest national park in the country and is known for its beautiful forests, rivers, and lakes.

Latvia is famous for its rich and unique folk traditions, including traditional music, dance, and crafts. The country is also home to a number of annual folk festivals, including the Latvian Song and Dance Festival, which is held every five years and features thousands of performers from around the country.

Latvia has a rich and unusual architectural heritage, including a number of historic wooden buildings, including churches and farmhouses. Many of these buildings are now protected as cultural landmarks and are popular tourist destinations.

Latvia is known for its beautiful and unusual sand dunes, which are formed by wind blowing sand from the beach into the interior of the country. The largest sand dunes in Latvia are located in the city of Jūrmala and are a popular destination for outdoor enthusiasts.

Latvia has a rich and unusual history, including a long period of foreign domination, including a period of German, Polish, and Swedish rule. The country was also occupied by the Soviet Union during the 20th century and regained its independence in 1991. Today, Latvia is a member of the European Union and NATO and is recognized as a modern, democratic state with a rich cultural heritage.

Liechtenstein

Liechtenstein is the sixth smallest country in the world and is located between Switzerland and Austria. It has a population of only around 38,000 people.

Liechtenstein is one of the only two doubly landlocked countries in the world, meaning that it is a landlocked country within a landlocked country.

Liechtenstein is known for its strong economy and is one of the wealthiest countries in the world, with a high standard of living and a low unemployment rate.

Liechtenstein has a long history of political neutrality and has never been involved in a war. The country is also known for its strong tradition of direct democracy, with citizens having the ability to petition for a national referendum on any issue.

Despite its small size, Liechtenstein has a rich cultural heritage and is known for its beautiful architecture, including historic castles and churches. The country is also home to several museums, including the Liechtenstein National Museum, which showcases the country's history and cultural heritage.

Lithuania

Lithuania is home to several unusual and unique natural landscapes, including the Curonian Spit, a narrow strip of land that separates the Baltic Sea from the Curonian Lagoon. The Curonian Spit is a UNESCO World Heritage Site and is known for its beautiful sand dunes, forests, and lakes.

Lithuania is famous for its rich and unusual folk traditions, including traditional music, dance, and crafts. The country is also home to a number of annual folk festivals, including the Vilnius International Folklore Festival, which is held every summer and features performances by folk musicians and dancers from around the world.

Lithuania is one of the few countries in Europe to have a high number of endangered species, including the European bison, lynx, and wolf. The country is also home to several unique and unusual bird species, including the black stork and the white-tailed eagle.

Lithuania is known for its rich and unusual history, including a long period of foreign domination, including a period of Polish, Russian, and Soviet rule. The country regained its independence in 1990 and is now a member of the European Union and NATO.

Despite its small size, Lithuania has a rich cultural heritage and is known for its beautiful architecture, including historic churches and castles. The country is also home to several museums, including the Lithuanian National Museum, which showcases the country's history and cultural heritage.

Luxembourg

Luxembourg is one of the smallest countries in Europe and is located in the heart of the continent, surrounded by Belgium, Germany, and France.

Despite its small size, Luxembourg is one of the wealthiest countries in the world, with a strong economy and a high standard of living.

Luxembourg is a multicultural country, with a large expatriate population and a rich history of immigration. The country is known for its diverse cultural heritage, including its traditional food, music, and art.

Luxembourg is home to several unusual and unique natural landscapes, including the Ardennes Forest, which covers a large portion of the country and is known for its scenic beauty and hiking trails.

Luxembourg has a long and rich history, including a period of independence, followed by centuries of foreign domination, including periods of French, Austrian, and German rule. Today, Luxembourg is a constitutional monarchy and a member of the European Union and NATO.

Malta

Malta is a small island nation located in the Mediterranean Sea and is known for its rich cultural heritage, warm climate, and stunning natural landscapes.

Malta has a long and rich history, including periods of Phoenician, Roman, and Arab rule, which have all left their mark on the country's culture, architecture, and language.

Malta is known for its unusual and unique architecture, including the Megalithic Temples of Malta, which are some of the oldest freestanding structures in the world and date back to over 5,000 years ago.

Malta is a popular tourist destination and is known for its stunning beaches, crystal-clear waters, and picturesque villages. The country is also home to several historic fortifications, including Valletta, which was built in the 16th century to defend against Ottoman invasions.

Malta is a unique and unusual place to visit, with a rich cultural heritage, stunning natural landscapes, and a warm and welcoming people. Whether you're interested in history, nature, or just soaking up the sun, Malta has something to offer for everyone.

Moldova

Moldova is a landlocked country located in Eastern Europe and is bordered by Romania and Ukraine. Despite its small size, Moldova has a rich cultural heritage and is known for its diverse landscape, including rolling hills, forests, and vineyards.

Moldova is the largest wine producer in Europe per capita, and the country's wine industry dates back to the ancient times of the Roman Empire. In fact, Moldovan wine is considered by many to be among the best in the world.

Moldova is a unique blend of Eastern European and Mediterranean cultures, with a rich history that includes periods of Ottoman, Russian, and Soviet rule. This cultural heritage is reflected in the country's diverse architecture, cuisine, and traditions.

The capital city of Chisinau is known for its unusual mix of Soviet-era architecture and modern buildings, as well as its parks and gardens, which are among the largest and most beautiful in Eastern Europe.

Moldova is home to several unusual and unique festivals and celebrations, including the annual Wine Festival, which takes place in the town of Cricova and features music, dance, and of course, a lot of wine. The country is also famous for its traditional Easter celebrations, which include singing, feasting, and the giving of painted eggs.

Monaco

Monaco is a tiny city-state located on the French Riviera and is known for its luxurious lifestyle, beautiful coastline, and warm climate. Despite its small size, Monaco has a rich cultural heritage and a long history dating back to the 12th century.

Monaco is famous for its high-end shopping, fine dining, and world-class casinos, making it a popular destination for the rich and famous. The country is also home to several beautiful and historic palaces, including the Prince's Palace of Monaco, which has been the official residence of the ruling monarchs of Monaco for centuries.

Monaco has the highest police presence per capita of any country in the world, ensuring that the streets are always safe and secure for its residents and visitors.

Monaco is known for its beautiful and unusual gardens, including the Jardin Exotique, which is located on a cliff overlooking the

Mediterranean Sea and features a collection of cacti, succulents, and other exotic plants.

Despite its small size, Monaco is a country of surprising contrasts, with modern and bustling cities, beautiful natural landscapes, and a rich cultural heritage. Whether you're interested in history, culture, or just soaking up the sun, Monaco has something to offer for everyone.

Montenegro

Montenegro is a small country located in the Balkans and is known for its stunning natural beauty, including mountains, beaches, and historic cities.

The country has a rich cultural heritage and a long history, dating back to the medieval era. This history is reflected in the country's many historical landmarks and monuments, including medieval castles, monasteries, and churches.

Montenegro is home to Lake Skadar, which is the largest lake in the Balkans and is surrounded by scenic mountains and rolling hills. The lake is also an important wildlife habitat, and is home to a variety of bird species, including pelicans, herons, and cormorants.

The country is known for its strong maritime tradition, and is home to a number of historic ports and harbors, including the city of Kotor, which is located on the coast of the Bay of Kotor and is surrounded by steep mountains and fortifications.

Montenegro is also famous for its unusual and unique customs and traditions, including the Pust Festival, which takes place during the Christian season of Lent and features elaborate masks, costumes, and dances. Another popular tradition is the Kokanje, a traditional horseback riding competition that takes place in the rural areas of the country and is steeped in folklore and tradition.

Netherlands

The Netherlands is known for its flat landscape, which was created through centuries of intensive land reclamation and dike building. Despite its flat terrain, the country is home to many beautiful and scenic landscapes, including rolling hills, vast wetlands, and the famous tulip fields.

The Dutch are famous for their love of cheese, and the country is home to over 200 different types of cheese, including Gouda, Edam, and Leiden. In addition, the Netherlands is one of the largest exporters of cheese in the world.

The Netherlands is also known for its extensive network of canals, which are an important part of the country's transportation system and also provide a picturesque backdrop to many of the country's cities.

The Dutch are famous for their love of bicycles, and the country is one of the most bicycle-friendly in the world, with extensive bike lanes and dedicated bicycle paths throughout the country.

The Netherlands is also home to a number of unusual and quirky museums, including the Museum of Bags and Purses, the Dutch Cheese Museum, and the Sex Museum. In addition, the country is famous for its elaborate and colorful flower festivals, including the Keukenhof Gardens and the annual Flower Parade in Haarlem.

North Macedonia

North Macedonia is a small country located in the Balkans region and is known for its rich history and cultural heritage. The country was once part of the ancient Kingdom of Paeonia, and has been inhabited by many different civilizations over the centuries, including the Illyrians, Romans, and Ottoman Turks.

North Macedonia is home to a number of unique and beautiful landscapes, including the stunning Lake Ohrid, which is one of the oldest and deepest lakes in Europe and is surrounded by scenic mountains and rolling hills.

The country is famous for its traditional crafts and artisanal products, including hand-woven textiles, pottery, and metalwork. Many of these traditional crafts are still made using traditional techniques that have been passed down from generation to generation.

North Macedonia is also known for its rich musical tradition, and is home to a number of different styles of music, including traditional folk music, pop, and classical music. The country is also famous for its lively festivals and events, including the Ohrid Summer Festival, which is a celebration of music, dance, and culture.

One of the most unusual and unique features of North Macedonia is its cuisine, which is a fusion of Balkan and Mediterranean flavors and ingredients. The country is famous for its hearty stews and grilled meats, as well as its delicious pastries and sweets, including baklava and tulumbe, which are made with phyllo dough and a sweet syrup or cream filling.

Norway

Norway is known for its stunning natural landscapes, including the fjords, mountains, and glaciers that make up much of the country's terrain. One of the most unusual and unique landscapes in Norway is the Lofoten Islands, which are located above the Arctic Circle and are known for their dramatic scenery and unusual rock formations.

Norway has a rich and varied folklore tradition, and is home to many legends and tales about trolls, elves, and other supernatural creatures. One of the most famous legends is that of the Huldra, a beautiful and mysterious forest creature said to live in the forests and mountains of Norway.

Norway is one of the wealthiest and most technologically advanced countries in the world, and is known for its high standard of living and its progressive policies, including its strong focus on environmental protection and sustainability.

Despite its modern and advanced society, Norway is still home to many traditional and rural communities, where life is much simpler and people still rely on fishing, farming, and hunting for their livelihoods.

One of the most unusual aspects of life in Norway is the long hours of darkness that are experienced during the winter months in many parts of the country. However, this darkness is offset by the beautiful and ethereal Northern Lights, which can often be seen dancing across the night sky in a mesmerizing display of colors and light.

Poland

Poland is known for its rich and diverse history, and has a long and fascinating cultural heritage that dates back many centuries.

One of the most unusual and unique aspects of Polish history is the tradition of Paper Cutting, which is a form of folk art that involves creating intricate and detailed designs from paper.

Poland is home to many unique and unusual celebrations, including the traditional Dozynki Harvest Festival, which is a celebration of the end of the harvest season and the beginning of winter. During the festival, people wear traditional clothing, dance, and sing, and there is also a parade of decorated carts and animals.

Poland is famous for its delicious cuisine, which is hearty and filling, and made from locally sourced ingredients. One of the most unusual and unique dishes in Polish cuisine is Bigos, which is a traditional stew made from cabbage, meat, and a variety of other ingredients.

Despite its modern and rapidly developing cities, Poland is still home to many rural and traditional communities, where life is much simpler and people still rely on farming, fishing, and hunting for their livelihoods.

One of the most unusual and unique features of the Polish landscape is the Bieszczady Mountains, which are located in the southeast of the country and are known for their stunning scenery, unspoiled forests, and diverse wildlife. The mountains are a popular destination for hikers, nature lovers, and those seeking to escape the hustle and bustle of modern life.

Portugal

Portugal has a long and rich history, and is known for its cultural heritage and many unique traditions. One of the most unusual and interesting aspects of Portuguese culture is the tradition of Fado music, which is a form of musical storytelling that is often emotional and melancholic in nature.

Portugal is famous for its delicious cuisine, which is rich and flavorful, and made from locally sourced ingredients. One of the most unusual and unique dishes in Portuguese cuisine is the traditional dish of Bacalhau, which is a salt cod dish that is often served with potatoes, onions, and eggs.

Portugal is home to many beautiful and historic cities, including the capital city of Lisbon, which is known for its winding streets, charming squares, and historic monuments. Lisbon is also home to many unique and unusual attractions, including the Elevador de Santa Justa, which is a historic elevator that offers panoramic views of the city.

Portugal is also home to many beautiful and unspoiled beaches, which are popular destinations for tourists and locals alike. One of the most unusual and unique beaches in Portugal is Praia da Ursa, which is a secluded and remote beach located in the Sintra-Cascais Natural Park.

Portugal is famous for its rich history, and is home to many historic monuments and landmarks, including the famous Torre de Belém, which is a 16th century tower that was built to commemorate Portugal's Age of Discovery. The tower is now a UNESCO World Heritage Site and is considered one of the most important symbols of Portuguese history and cultural heritage.

Romania

Romania is famous for its rich history and cultural heritage, and is home to many unique and unusual traditions and customs. One of the most interesting aspects of Romanian culture is the tradition of

Maramureş wooden churches, which are wooden structures that are carved and decorated in intricate and elaborate designs.

Romania is also famous for its stunning natural landscapes, including the Carpathian Mountains, which are home to many unique and unusual flora and fauna, including wolves, bears, and lynx. The mountains are also home to the famous Bâlea Lake, which is a glacial lake located at an altitude of over 2,000 meters, and is one of the most picturesque and remote destinations in Romania.

Romania is also home to many historic castles and fortresses, including the famous Corvin Castle, which is considered one of the most beautiful and well-preserved castles in Europe. The castle is known for its stunning architecture and rich history, and is now a popular tourist destination in Romania.

Romania is also famous for its rich and vibrant folklore, which is full of legends, myths, and stories of supernatural beings and creatures. One of the most famous and well-known legends in Romanian folklore is the story of the vampire Count Dracula, which is said to be based on the historical figure of Vlad III Dracula, who was a prince of Wallachia in the 15th century.

Romania is also famous for its rich musical heritage, and is home to many unique and unusual musical traditions, including the pan flute, which is a traditional musical instrument that is made from bamboo and is played by blowing into the end of the tubes. The pan flute is considered one of the most iconic symbols of Romanian culture, and is featured in many traditional Romanian songs and folk dances.

Russia

Russia is home to Lake Baikal, the world's deepest and largest freshwater lake. It is also one of the world's clearest and purest lakes, with visibility that can reach up to 40 meters. Lake Baikal is considered one of the most unique and unusual natural wonders in the world, and is home to many rare and endemic species of plants and animals.

Russia is famous for its grand and impressive architecture, including St. Basil's Cathedral in Moscow, which is a unique and unusual building that was built in the 16th century and is considered one of the most iconic landmarks of Russia.

Russia is also home to the famous Hermitage Museum in St. Petersburg, which is one of the largest and oldest art museums in the world, and is home to an enormous collection of art and artifacts from many different cultures and periods of history.

Russia is famous for its rich and vibrant folk traditions, including the traditional Russian matryoshka dolls, which are nested dolls that are carved and painted in elaborate and colorful designs. The dolls are considered one of the most iconic symbols of Russian folk art, and are popular souvenirs and gifts for tourists visiting Russia.

Russia is also famous for its long and harsh winters, which can last for several months and can be extremely cold and snowy. Despite this, many Russians embrace winter and participate in traditional winter activities such as ice skating, sledding, and ice fishing. One of the most unusual winter traditions in Russia is the winter bathing, where people go swimming in the cold lakes and rivers, which is believed to be good for health and well-being.

San Marino

San Marino is one of the world's smallest countries, with a total area of only 61 square kilometers and a population of just over 33,000 people. Despite its small size, it has a rich and fascinating history, dating back to the year 301 AD when it was founded as a republic.

San Marino is one of the world's oldest republics, and is often referred to as the oldest surviving sovereign state and constitutional republic. It has a unique system of government, with two captains regent serving as the country's head of state for six-month terms.

San Marino is known for its stunning natural beauty, including the stunning Monte Titano, a mountain that serves as the symbol of the country and is featured on its national flag. The mountain offers panoramic views of the surrounding countryside and is a popular tourist destination.

San Marino is home to a number of historic buildings and landmarks, including the Three Towers of San Marino, which are a series of three fortresses located on the mountain and offer stunning views of the surrounding landscape.

San Marino is also famous for its postal stamps, which are highly collectible and are known for their intricate and beautiful designs. The country produces a wide variety of stamps on many different themes, and is considered one of the world's leading producers of philatelic items. San Marino stamps are a unique and unusual collectible, and are highly prized by stamp collectors around the world.

Serbia

Serbia is one of the world's largest exporters of raspberries. The country's fertile soil and warm climate make it ideal for growing the fruit, and it is estimated that around 60% of the world's raspberries come from Serbia.

Serbia is known for its vibrant and diverse music scene, which includes a variety of traditional folk music styles as well as popular contemporary music. The country is also home to many famous composers and musicians, including the famous composer Stanislav Binicki, who is known for his contributions to Serbian classical music.

The city of Belgrade, the capital of Serbia, is built on the confluence of the Sava and Danube rivers, making it one of the few cities in the world located at the crossroads of two major rivers.

Serbia is home to a number of unique festivals and events, including the Guča Trumpet Festival, which is one of the largest brass music festivals in the world and attracts thousands of visitors each year. The festival features performances by talented brass bands from around the world and is known for its lively atmosphere and celebration of traditional Serbian music and culture.

Serbia is also famous for its delicious cuisine, which is a fusion of Turkish, Hungarian, and Austrian influences. Some of the country's most famous dishes include cevapi (grilled meat skewers), gibanica (a savory pastry made with cheese and phyllo dough), and ajvar (a roasted red pepper spread). Serbian cuisine is considered some of the tastiest in Europe, and is well worth trying if you ever have the opportunity.

Slovakia

Slovakia has the highest number of castles and chateaus per capita in the world, with over 200 well-preserved castles and chateaus scattered throughout the country. Some of the most famous include the Spiš Castle, the Bratislava Castle, and the Orava Castle.

Slovakia is known for its traditional folk art, which is influenced by the country's rich history and cultural heritage. One of the most famous forms of folk art in Slovakia is the wooden carved sculptures, known as the "gombocy," which are used to decorate houses and other buildings.

The Slovak National Uprising, which took place in 1944, was one of the largest resistance movements against Nazi rule during World War II. The uprising was led by Slovak soldiers and civilians and was an important turning point in the war, helping to pave the way for the eventual defeat of the Nazis.

Slovakia is home to the Tatra Mountains, which are part of the Carpathian Mountain range and are considered some of the most beautiful mountain landscapes in Europe. The Tatra Mountains are a popular destination for outdoor enthusiasts, offering a range of activities including hiking, skiing, and rock climbing.

Slovakia is known for its thriving music scene, with a rich tradition of folk music, classical music, and popular music. The country is also home to a number of music festivals, including the Slovak National Music Festival and the Pohoda Festival, which is one of the largest music festivals in Central Europe and features a diverse lineup of international and local musicians.

Slovenia

Slovenia is the only country in the world that is fully lit by green energy. The country has invested heavily in renewable energy sources, such as hydropower and wind power, and generates 100% of its electricity from clean energy sources.

Slovenia is one of the few countries in the world that has a natural phenomenon called "human fish." This refers to the blind, subterranean salamanders that live in the caves of Slovenia and are only found in this part of the world.

Slovenia is known for its vibrant and diverse food culture, which has been influenced by its geographic location and historical ties to neighboring countries. Some of the most popular dishes in Slovenia include Carniolan sausage, potica (a sweet bread), and the traditional "potica" cake.

Slovenia is home to the Julian Alps, which are part of the Eastern Alps and are considered some of the most beautiful mountain landscapes in Europe. The Julian Alps are a popular destination for outdoor enthusiasts, offering a range of activities including hiking, skiing, and rock climbing.

Slovenia is famous for its rich wine-making tradition, with a long history of producing high-quality wines. Some of the most famous wine regions in Slovenia include the Goriska Brda and the Vipava Valley, which are known for producing some of the best white wines in the country.

Spain

Spain is home to the longest coastline in Europe, stretching over 4,000 kilometers along the Mediterranean and Atlantic Oceans.

Spain is the world's largest producer of olive oil, producing over 50% of the world's olive oil each year.

Spain is known for its bullfighting culture, which has been a tradition in the country for over 2,000 years. Bullfighting is considered an art form in Spain, and the bullfighters are seen as heroes.

Spain has the largest number of protected natural areas in Europe, including 14 national parks and numerous natural reserves. Some of the most famous protected areas include the Doñana National Park and the Pyrenees Mountains.

The famous Sagrada Familia in Barcelona is one of the most visited tourist attractions in Spain. It is a large, unfinished church that has been under construction since 1882 and is expected to be completed in 2026, over 140 years after it first began.

Sweden

Sweden is home to the largest number of moose in Europe, with an estimated population of around 400,000 moose.

Sweden is one of the world's most environmentally conscious countries, with a focus on sustainability and renewable energy. Over 50% of the country's energy is produced from renewable sources.

In Sweden, it is customary to take off your shoes when entering someone's home. This is to keep the floors clean and to show respect for the homeowner.

Sweden is known for its Midsummer festival, which is celebrated on the longest day of the year (around June 21st). During the festival, people dance around a Maypole, eat traditional Swedish food, and celebrate the arrival of summer.

Sweden has a high level of gender equality, with one of the smallest gender pay gaps in the world. The country also has some of the most progressive parental leave policies, allowing parents to take up to 480 days off work after the birth of a child, with 80% of their pay covered by the government.

Switzerland

Switzerland has four official languages: German, French, Italian, and Romansh. This diversity is a reflection of the country's unique cultural and linguistic heritage.

Switzerland is known for its banking industry and strict privacy laws. It is illegal for Swiss banks to reveal information about their clients, even to the government.

Switzerland is home to the world's largest particle physics laboratory, the Large Hadron Collider (LHC), which is located underground near Geneva.

Switzerland has a long tradition of neutrality and has not participated in a war since 1815. The country has served as a neutral host for international peace negotiations, including the talks that ended the First World War.

Switzerland has a rich history of clock-making, with some of the world's most intricate and complex timepieces originating from the country. In the city of Geneva, there is a famous watchmaking school, the Ecole d'Horlogerie de Genève, that trains future generations of watchmakers.

Ukraine

Ukraine is the second-largest country in Europe, covering an area of 603,550 square kilometers. Despite its large size, Ukraine has a relatively low population density, with most of its residents concentrated in urban areas.

Ukraine has a rich and diverse cultural heritage, with a long history of folk art, music, and dance. The country is home to numerous historical sites and cultural landmarks, including the UNESCO World Heritage site of Kyiv-Pechersk Lavra.

Ukraine is one of the world's leading producers of sunflower seeds, wheat, and corn. The country's fertile farmland and favorable climate make it an ideal location for agriculture.

Ukraine is home to the world's largest disco ball, which is located in the city of Dnipro. The giant ball, which weighs over 5 tons, is a popular tourist attraction and symbol of the city's vibrant cultural scene.

Ukraine has a long and rich history of science and technology, with many Ukrainian scientists and engineers making important contributions to fields such as mathematics, physics, and space exploration. One of the country's most famous sons is the cosmonaut Leonid Kadenyuk, who was the first Ukrainian to fly in space.

United Kingdom

The UK is home to a number of unusual and quirky sporting events, including the "World Gravy Wrestling Championships," which takes place annually in Lancashire and involves competitors wrestling in a vat of gravy.

The UK is home to a number of eccentric traditions, including "Cheese Rolling," which is held annually in Gloucestershire and involves competitors racing down a steep hill after a wheel of cheese.

The UK is the birthplace of many bizarre and unusual musical genres, including "Nerdcore Hip Hop," a subgenre of hip hop music that focuses on geek and nerd culture, and "Chap-Hop," a genre of music that fuses elements of hip hop and English folk music.

In the UK, there is a tradition of "well-dressing," in which communities decorate public wells and springs with elaborate designs made from flowers and other natural materials. This tradition is thought to date back to the 16th century and is still observed in some parts of the country today.

The UK is home to a number of unusual museums, including the "International Spoon Museum," which is located in Devon and contains over 5,000 spoons from around the world, and the "British Lawnmower Museum," which is located in Southport and contains over 200 vintage and antique lawnmowers.

Vatican City

Vatican City (sovereign city-state located within the city of Rome, Italy and it is considered the smallest country in Europe. It is an independent, landlocked enclave surrounded by the city of Rome and is the headquarters of the Roman Catholic Church. Vatican City is governed as an absolute monarchy with the Pope as its head of state. It has its own government).

Vatican City is the smallest independent country in the world, both in terms of size and population. It covers an area of just 44 hectares and has a population of around 800 people.

Home to St. Peter's Basilica: St. Peter's Basilica is located in Vatican City and is considered one of the largest and most important churches in the world. It is the burial site of Saint Peter, one of the twelve apostles of Jesus.

Vatican City is home to many of the world's greatest works of art, including sculptures, paintings, and mosaics. It is one of the most important art museums in the world, with many priceless masterpieces on display.

The Vatican's official army is the Swiss Guard, a small force of Swiss soldiers who have been responsible for the security of the pope for over 500 years.

Vatican City is the only country in the world where citizens don't have to pay taxes. The cost of running the country is mostly covered by donations and the sale of souvenirs and postage stamps.

AFRICA

Africa is the second-largest and second most populous continent in the world, with a population of over 1.3 billion people. It is located in the Northern and Western hemispheres and is bordered by the Mediterranean Sea to the north, the Red Sea to the northeast, the Indian Ocean to the southeast, and the Atlantic Ocean to the west.

The African continent is rich in natural resources, including minerals, oil, and natural gas. It is also home to some of the world's most diverse ecosystems and is home to many of the world's most endangered species.

Africa has a rich cultural heritage, with a diverse range of languages, religions, and traditions. The continent is also known for its art, music, and traditional crafts, and for its long history of tribal and ethnic conflict. Despite these conflicts, Africa is also known for its hospitality and friendliness.

In recent years, Africa has experienced rapid economic growth, with many countries experiencing strong increases in foreign investment and tourism. However, the continent continues to face many challenges, including poverty, disease, and conflict, and many countries struggle with poor governance and lack of infrastructure. Nevertheless, Africa is a continent of immense potential, with many countries poised to play a major role in shaping the future of the world.

There are 54 recognized sovereign states in Africa that are members of the United Nations (UN). All of these countries are recognized as sovereign states and have a seat at the UN General Assembly, where they can participate in debates, negotiations, and decision-making processes affecting the international community.

Algeria

Algeria has a large underground oasis, known as the Tassili n'Ajjer National Park, which is home to over 15,000 prehistoric rock paintings.

The city of Algiers is built on the slopes of a hill, and its traditional architecture features distinctive white and blue buildings.

Algeria is one of the few countries in the world that has both a Mediterranean coastline and a desert interior, giving it a unique and diverse landscape.

The Algerian flag features a unique emblem known as the "Star of Algeria", which represents the country's hope for freedom and independence.

Algeria is known for its traditional cuisine, which features a variety of dishes made from locally grown ingredients, such as dates, figs, and olives. One of the most famous dishes is couscous, a staple food made from semolina grains.

Angola

Angola has a significant Portuguese influence, as it was a former Portuguese colony. Portuguese is still widely spoken in the country and is one of its official languages.

The capital city of Luanda is one of the most expensive cities in the world, with a cost of living that rivals cities like Tokyo and London.

Angola is home to a species of giant sengi, also known as the elephant-shrew, which can grow up to 14 inches long and is one of the fastest land animals in the world.

The country has a rich and diverse culture, with over 80 ethnic groups, each with its own distinct customs, music, and cuisine.

Angola is rich in natural resources, including diamonds, oil, and gold, but despite this, it remains one of the poorest countries in the world, with widespread poverty and limited access to basic services like healthcare and education.

Benin

Benin is known as the birthplace of the voodoo religion, which is still widely practiced in the country today and is an important part of its cultural heritage.

The city of Ouidah, located in western Benin, was a major center for the slave trade in the 17th and 18th centuries and is home to the famous "Route des Esclaves," or "Slave Route," a historical monument that commemorates the journey of enslaved Africans.

Benin has a rich history of traditional African art and crafts, particularly in the form of bronze and brass sculptures, which were created by the Kingdom of Dahomey in the 16th century and are now renowned for their beauty and historical significance.

Benin is home to the largest mud mosque in the world, the Great Mosque of Ouidah, which is made entirely of mud and was built in the 19th century.

The country is located on the Gulf of Guinea and is part of the West African coastline, which is known for its diverse and abundant marine life, including sea turtles, whales, and dolphins.

The Parc National de la Pendjari in northern Benin is a popular destination for wildlife viewing and is home to many species of rare and endangered animals, including the West African lion.

Botswana

Botswana is home to the largest population of elephants in the world, and the country has made a commitment to conservation, with over a third of its land set aside for national parks and wildlife reserves.

The country has one of the fastest-growing economies in the world, due in large part to its abundant diamond reserves and a thriving tourism industry.

Botswana is known for its unique cultural traditions, including the practice of "kgotla," a form of community decision-making where villagers gather to discuss and resolve issues collectively.

The Kalahari Desert, which covers a large portion of Botswana, is known for its unique flora and fauna, including the rare and endangered African wild dog.

Despite its rapid economic growth, Botswana has a long-standing commitment to social and economic equality, and is one of the few countries in the world with a high Human Development Index and a low income inequality index.

Burkina Faso

Burkina Faso is one of the poorest countries in the world, with a large portion of its population relying on subsistence agriculture for their livelihoods.

The country has a rich cultural heritage, with a long history of traditional arts and crafts, including pottery, weaving, and metalworking.

Burkina Faso is home to a large number of indigenous languages, with over 60 ethnic groups and more than 60 different languages spoken throughout the country.

The capital city of Ouagadougou is known for its vibrant cultural scene, including music, dance, and theater, as well as its thriving arts and crafts markets.

Burkina Faso is located in the Sahel region of West Africa, which is known for its harsh and unpredictable climate, with long dry seasons and occasional flash floods. Despite these challenges, the people of Burkina Faso have a strong cultural tradition of resilience and resourcefulness.

Burundi

Burundi is one of the smallest countries in Africa, with a population of over 11 million people. Despite its small size, it is one of the most densely populated countries on the continent.

The country is located in the Great Lakes region of Africa and is home to several large lakes, including Lake Tanganyika, which is one of the deepest and oldest lakes in the world.

Burundi has a long history of political instability and conflict, including a 12-year civil war that ended in 2005. Despite these challenges, the country has made significant progress in recent years towards peace and stability.

Burundi is known for its traditional music and dance, which are an important part of its cultural heritage and are performed at festivals and celebrations throughout the country.

The country is located in the African Rift Valley, which is home to a large number of active and dormant volcanoes, as well as hot springs and geysers. The area is known for its unique geothermal features and is a popular tourist destination.

Cabo Verde

Cabo Verde, also known as Cape Verde, is an island nation located off the west coast of Africa. The country is made up of ten volcanic islands, nine of which are inhabited.

Cabo Verde has a rich cultural heritage, including music, dance, and literature, that is heavily influenced by African, Portuguese, and Brazilian traditions. The country is especially known for its distinctive style of music called morna, which is a form of slow, soulful ballad that often tells stories of love, loss, and the struggles of life.

Despite its small size, Cabo Verde has a high level of biodiversity, with a wide range of plant and animal species, many of which are unique to the country. The islands are home to several protected areas, including the Monte Gordo Natural Park, which is a popular destination for wildlife viewing.

Cabo Verde is a former Portuguese colony and has a strong historical connection to Brazil, as many Cabo Verdeans emigrated to Brazil in search of work in the 19th and 20th centuries. This connection is reflected in the country's language, cuisine, and cultural traditions.

The country is known for its beautiful beaches and clear waters, making it a popular destination for beach vacations and water sports. The island of Boa Vista is especially known for its long, white sand beaches and clear waters, which are perfect for swimming, snorkeling, and scuba diving.

Cameroon

Cameroon is located in West Africa and is known for its diverse geography, which includes everything from tropical forests and deserts to mountain ranges and savannas. The country is sometimes referred to as "Africa in miniature" due to its varied landscapes and cultures.

Cameroon is home to over 250 ethnic groups, each with its own unique language, traditions, and customs. The country is also home to several indigenous pygmy communities, who have lived in the forested regions of Cameroon for thousands of years.

Cameroon is one of the largest producers of cocoa in the world, and its chocolate is considered some of the best in the world. The country is also a major producer of coffee, rubber, and oil.

Cameroon is home to several unique species of wildlife, including gorillas, chimpanzees, and elephants, as well as a wide variety of birds and reptiles. The country is also home to several national parks and wildlife reserves, including the Waza National Park, which is one of the largest wildlife reserves in West Africa.

The country is known for its rich cultural heritage, including music, dance, and traditional crafts, such as weaving and pottery. The city of Yaoundé is home to several museums and cultural centers, including the National Museum of Cameroon, which showcases the country's rich history and cultural heritage.

Central African Republic

The Central African Republic is a landlocked country located in the heart of Africa, bordered by Chad, Sudan, South Sudan, the Democratic Republic of the Congo, the Republic of Congo, and Cameroon. The country is known for its diverse landscape, including savannas, forests, and mountains.

The Central African Republic is rich in minerals, including diamonds, gold, and uranium, and is one of the largest producers of diamonds in Africa. Despite its mineral wealth, the country remains one of the poorest in the world, with a per capita income of less than $1,000.

The Central African Republic is home to several unique species of wildlife, including elephants, gorillas, and several species of primates. The country is also home to several national parks and wildlife reserves, including the Manovo-Gounda St. Floris National Park, which is a UNESCO World Heritage Site.

The country has a rich cultural heritage, with over 80 ethnic groups and a long history of traditional music, dance, and storytelling. The country is also known for its traditional crafts, including pottery, weaving, and metalworking.

Despite its rich cultural heritage and natural resources, the Central African Republic has experienced years of conflict and political instability, including a civil war that lasted from 2013 to 2020. Despite these challenges, the country has made significant progress in recent years towards peace and stability, and is working to rebuild its economy and infrastructure.

Chad

Chad is a landlocked country located in North-Central Africa, bordered by Libya, Sudan, the Central African Republic, Cameroon, Nigeria, and Niger. It is the fifth-largest country in Africa and one of the least densely populated.

Chad is home to several unique species of wildlife, including elephants, lions, cheetahs, and several species of primates. The country is also home to several national parks and wildlife reserves, including the Zakouma National Park, which is one of the largest wildlife reserves in Africa.

Despite its rich natural resources, including oil, gold, and uranium, Chad is one of the poorest countries in the world, with a per capita income of less than $1,000. The country has a long history of conflict and political instability, which has impacted its ability to develop its economy and infrastructure.

Chad is known for its diverse cultural heritage, with over 200 ethnic groups and a long history of traditional music, dance, and storytelling. The country is also known for its traditional crafts, including pottery, weaving, and metalworking.

Chad is home to Lake Chad, one of the largest freshwater lakes in Africa. The lake, which straddles the borders of Chad, Nigeria, Cameroon, and Niger, is a vital source of water for the region and supports a rich diversity of fish and other aquatic species. In recent decades, the lake has shrunk dramatically due to a combination of factors, including climate change, over-extraction of water for irrigation, and population growth.

Comoros

The Comoros is an island nation located in the Indian Ocean, off the coast of East Africa. The country consists of four main islands

- Grande Comore, Moheli, Anjouan, and Mayotte - as well as several smaller islands.

The Comoros is known for its rich cultural heritage, with a long history of traditional music, dance, and storytelling. The country is also known for its unique cuisine, which combines African, Arab, and French influences.

The Comoros is one of the world's poorest countries, with a per capita income of less than $1,000. Despite its poverty, the country is rich in natural resources, including fertile soil, abundant rainfall, and a warm tropical climate that supports a wide variety of crops and livestock.

The Comoros is home to several unique species of wildlife, including lemurs, bushbabies, and several species of birds. The country is also home to several national parks and wildlife reserves, including the Halle aux Poissons Reserve, which is a popular destination for snorkeling and diving.

The Comoros has a rich history of maritime trade and exploration, and was a major center of the slave trade during the 19th century. The country is also known for its vibrant independence movement, which has seen the Comoros gain independence from France in 1975, and its ongoing efforts to develop its economy and infrastructure.

There are two countries called the Democratic Republic of the Congo (DRC) and the Republic of the Congo (ROC), also known as Congo-Brazzaville, which are located in Central Africa. The DRC is located in the larger and more densely populated part of the Congo region and is the third largest country in Africa by area. The ROC is a smaller, coastal country located to the west of the DRC. The two countries are often confused because of their similar names and close proximity, but they are distinct and separate nations with different governments, economies, and cultures.

Democratic Republic of the Congo

The Democratic Republic of the Congo (DRC) is a large country located in Central Africa, bordered by nine other African countries. It is the second-largest country in Africa by land area and is home to over 100 million people, making it one of the most populous countries in Africa.

The DRC is known for its rich natural resources, including diamonds, gold, cobalt, and copper. The country is also home to some of the largest rainforests in the world, which are home to a rich diversity of plant and animal life, including gorillas, elephants, and many other species of primates.

Despite its abundant natural resources, the DRC is one of the poorest countries in the world, with a per capita income of less than $400. The country has a long history of conflict, political instability, and corruption, which has impacted its ability to develop its economy and infrastructure.

The DRC is known for its vibrant music and dance scenes, which reflect the country's rich cultural heritage and diverse population. The country is also known for its traditional crafts, including textiles, basketry, and wood carving.

The DRC has a rich history of art and culture, and is home to several important cultural landmarks, including the Congo River and the spectacular waterfalls of the Congo Basin, which are among the largest in the world. The country is also home to several world-class museums, including the Museum of the Royal

Palace in Kinshasa, which houses a rich collection of art, cultural artifacts, and historical objects from across the Congo Basin.

Republic of the Congo

The Republic of the Congo, also known as Congo-Brazzaville, is a country located in Central Africa, bordered by the Democratic Republic of the Congo to the west, Gabon to the northwest, Cameroon to the north, and the Cabinda Enclave of Angola to the southwest.

The Republic of the Congo is known for its rich natural resources, including oil, timber, and minerals, which provide the country with a strong source of revenue. Despite its wealth, the country is still one of the poorest in the world, with a per capita income of less than $1,500.

The Republic of the Congo is home to several unique species of wildlife, including gorillas, chimpanzees, and many other species of primates, as well as elephants, lions, and other large mammals. The country is also home to several national parks and wildlife reserves, including the Nouabale-Ndoki National Park, which is known for its rich biodiversity.

The Republic of the Congo is known for its vibrant music and dance scenes, which reflect the country's rich cultural heritage and diverse population. The country is also known for its traditional crafts, including textiles, basketry, and wood carving.

The Republic of the Congo has a rich history of art and culture, and is home to several important cultural landmarks, including the iconic Basilica of Sainte-Anne de Brazzaville, which is one of the largest and most important Catholic churches in the country. The country is also home to several world-class museums, including the National Museum of Brazzaville, which houses a rich collection of art, cultural artifacts, and historical objects from across the Congo Basin.

Côte d'Ivoire

Côte d'Ivoire, also known as Ivory Coast, is a country located in West Africa, bordered by Liberia and Guinea to the west, Mali and Burkina Faso to the north, and Ghana to the east. The country is known for its diverse culture, rich history, and vibrant economy.

Côte d'Ivoire is one of the largest producers of cocoa in the world, and the country's cocoa industry plays a major role in its economy. The country is also a major producer of coffee, palm oil, and rubber, and is home to several large companies and industrial facilities.

Côte d'Ivoire is home to several unique species of wildlife, including elephants, monkeys, and many other species of primates, as well as lions, leopards, and other large mammals. The country is also home to several national parks and wildlife reserves, including the Comoé National Park, which is known for its rich biodiversity.

The music and dance scenes of Côte d'Ivoire are diverse and reflect the country's rich cultural heritage and diverse population. The country is known for its traditional instruments, including the balafon and the djembe, as well as its traditional music styles, such as zouglou and coupe-decale.

Côte d'Ivoire is known for its rich history and cultural heritage, and is home to several important cultural landmarks, including the Baoule mask, which is an iconic symbol of the country's traditional culture and is considered to be one of the most important cultural artifacts of West Africa. The country is also home to several world-class museums, including the National Museum of Abidjan, which houses a rich collection of art, cultural artifacts, and historical objects from across West Africa.

Djibouti

Djibouti is a small country located in the Horn of Africa, bordered by Eritrea to the north, Ethiopia to the west and south, and Somalia to the southeast. The country is known for its unique geography, including the Red Sea and the Gulf of Aden, and its strategic location as a major shipping and transit hub.

Djibouti is home to several unique species of wildlife, including baboons, hyenas, and many other species of primates, as well as leopards, cheetahs, and other large mammals. The country is also home to several national parks and wildlife reserves, including the Day Forest National Park, which is known for its rich biodiversity.

Djibouti is known for its vibrant music and dance scenes, which reflect the country's rich cultural heritage and diverse population. The country is also known for its traditional crafts, including textiles, basketry, and wood carving.

Djibouti has a rich history of art and culture, and is home to several important cultural landmarks, including the Great Rift Valley, which is one of the largest geological features in the world and is a major tourist attraction. The country is also home to several world-class museums, including the National Museum of Djibouti, which houses a rich collection of art, cultural artifacts, and historical objects from across the region.

Djibouti has a unique geography that includes the Red Sea and the Gulf of Aden, making it an important hub for shipping and transit.

The country is home to several major ports, including the Port of Djibouti, which is one of the busiest and most important ports in the region, and is a major hub for shipping, transit, and logistics.

Egypt

Egypt is a country located in North Africa, and is known for its rich history and cultural heritage, including the ancient civilization of the Pharaohs and the Pyramids of Giza. Egypt is also known for its unique geography, including the Nile River and the deserts of the Sahara.

The ancient Egyptians believed that the heart was the source of all emotions, thoughts, and memories, and not the brain. As a result, when mummifying their dead, they would leave the brain inside the body but remove the heart and preserve it separately.

The Great Sphinx of Giza, located near the Pyramids of Giza, is one of the largest and most iconic sculptures in the world. The Sphinx is a mythical creature with the body of a lion and the head of a human, and is believed to have been built during the reign of the Pharaoh Khafre in the fourth dynasty.

In ancient times, the Egyptians used a form of writing called hieroglyphics, which used pictures to represent words and ideas. This form of writing was deciphered in the 19th century, and has provided insight into the history, culture, and daily life of ancient Egypt.

The River Nile, which runs through Egypt and several other countries in the region, is the longest river in the world and has played a central role in the history and development of ancient and modern Egypt. The Nile has been used for transportation, agriculture, and as a source of water for thousands of years, and is considered to be one of the most important rivers in the world.

Equatorial Guinea

Equatorial Guinea is a small country located in Central Africa, and is known for its diverse geography, including both coastal areas and mountainous regions. The country is also known for its rich cultural heritage, which reflects its diverse population and history.

Equatorial Guinea was one of the first African countries to achieve independence from colonial rule, gaining independence from Spain in 1968. Despite this, the country has a history of political instability and has been ruled by a small group of individuals and their families for much of its modern history.

Equatorial Guinea is one of the smallest Spanish-speaking countries in the world, and is one of the few countries in Africa where Spanish is an official language. The country also has a number of indigenous languages, including Fang, Bubi, and others.

Equatorial Guinea is home to several unique species of wildlife, including gorillas, chimpanzees, and elephants, as well as a diverse range of bird species, including many species of parrots, kingfishers, and other birds. The country is also home to several national parks and wildlife reserves, including Monte Alen National Park and others.

Equatorial Guinea has a rich cultural heritage, which includes traditional dances, music, and festivals that reflect the country's diverse population and history. The country is also home to several important cultural landmarks, including the capital city of Malabo, which is known for its colonial architecture and rich history, and the island of Bioko, which is home to several important historic sites and cultural landmarks.

Eritrea

Eritrea is a country located in the Horn of Africa, and is known for its diverse geography, which includes highland regions, coastal areas, and deserts. The country is also known for its rich cultural heritage, which reflects its diverse population and history.

Eritrea has a long and complex history, including a long period of colonial rule by Italy and later by Ethiopia, followed by a brutal war for independence that lasted from 1961 to 1991. Despite this difficult history, Eritrea has emerged as a relatively stable and prosperous country in recent years.

Eritrea has a unique system of government, which is a combination of parliamentary democracy and presidential republic, with a strong emphasis on national unity and stability. The country has been ruled by President Isaias Afwerki since its independence in 1991.

Eritrea is one of the few countries in the world that does not have a single official religion, and instead recognizes the freedom of religion and the right of individuals to practice their own religion. This has helped to create a diverse and tolerant religious landscape in the country, which includes Christians, Muslims, and followers of traditional African religions.

Eritrea is home to several unique species of wildlife, including a variety of primates, birds, and reptiles, as well as several species of endemic plants and animals that are found nowhere else in the world. The country is also home to several important cultural landmarks, including the capital city of Asmara, which is known for its modernist architecture and rich history, and the region of Massawa, which is home to several important historic sites and cultural landmarks.

Eswatini (formerly Swaziland)

Eswatini is a small country located in southern Africa, and is known for its diverse geography, including highland regions, savannas, and forests. The country is also known for its rich cultural heritage, which reflects its diverse population and history.

Eswatini has a long and proud tradition of monarchy, and is one of the few remaining absolute monarchies in the world. The current king, King Mswati III, has ruled the country since 1986 and is known for his lavish lifestyle and support of traditional Swazi culture and customs.

Eswatini is one of the few countries in Africa that has managed to maintain a relatively high degree of political stability, despite its small size and the challenges faced by many other African countries. This stability has helped to promote economic growth and development in the country, and has made Eswatini one of the most prosperous countries in southern Africa.

Eswatini is known for its rich cultural heritage, which includes traditional music, dance, and festivals that reflect the country's diverse population and history. The country is also home to several important cultural landmarks, including the royal palace in Lobamba, which is the center of traditional Swazi culture and customs, and the famous Ezulwini Valley, which is home to several important cultural and religious sites.

Eswatini is home to several important conservation areas, including the Mkhaya Game Reserve and the Hlane Royal

National Park, which are both home to a diverse range of wildlife, including elephants, rhinos, and other species of large mammals. The country is also home to several species of endemic plants and animals, and is recognized as an important global center of biodiversity.

Ethiopia

Ethiopia is one of the world's oldest countries, with a history that dates back over 3,000 years. It is also considered one of the cradles of human civilization, as some of the earliest hominid fossils, including those of "Lucy," have been found there.

Ethiopia has its own calendar, which is based on the Coptic calendar and has 13 months, 12 of which have 30 days each and one with five or six extra days. The Ethiopian calendar is seven to eight years behind the Gregorian calendar, which is used in most of the world.

The city of Harar in Ethiopia is known for having a unique relationship with hyenas, which roam the streets and are fed by the local residents.

Ethiopia is the only African country that has never been fully colonized by a European power, although it was occupied by Italy from 1936 to 1941.

The Rastafarian movement has a strong connection to Ethiopia, particularly due to the influence of Haile Selassie I, the last emperor of Ethiopia. Selassie is considered by many Rastafarians to be the Messiah and the reincarnation of Jesus Christ. The Rastafarian movement, which originated in Jamaica in the 1930s, views Ethiopia as the spiritual homeland of African people and as a symbol of black pride and independence.

Gabon

Gabon is home to one of the largest remaining populations of lowland gorillas in the world. The gorillas are protected in the Ivindo National Park.

Gabon is one of the few countries in the world where it is illegal to cut down a tree, due to its commitment to preserving its lush rainforests and wildlife.

The country has a unique system of presidential succession, in which the Vice President automatically becomes the President if the incumbent President dies or resigns.

Gabon is one of the largest oil producers in Africa, and its oil reserves account for a significant portion of the country's GDP.

Gabon is known for its diverse and exotic wildlife, including elephants, lions, chimpanzees, and over 700 species of birds. The country has several national parks and reserves that protect its unique ecosystems and wildlife.

Gambia

Gambia is the smallest country in Africa: The Gambia is a small, narrow strip of land that runs along the banks of the Gambia River in West Africa. Despite its small size, The Gambia is one of the most densely populated countries in Africa.

Gambia is known for its rich bird life: The Gambia is home to a rich variety of bird species, making it a popular destination for birdwatchers. Some of the most commonly seen birds in The Gambia include flamingos, pelicans, storks, and many species of wading birds.

Gambia is a former British colony: The Gambia was a British colony from the 19th century until it gained independence in 1965. Today, The Gambia is a republic with a president and a parliamentary system of government.

Gambia is known for its beautiful beaches: The Gambia has a long, narrow coastline that is dotted with beautiful beaches and is popular with tourists. Some of the most popular beaches in The Gambia include Kotu Beach and Kololi Beach.

Gambia is a hub for traditional West African music: The Gambia is home to a vibrant music scene, with a rich tradition of traditional West African music and dance. Some of the most popular traditional music styles in The Gambia include griot music, a type of music that is performed by a caste of hereditary musicians, and kora music, which is played on a stringed instrument called a kora.

Ghana

Ghana is home to the largest man-made lake in the world: Lake Volta is the largest man-made lake in the world and is located in Ghana. It covers an area of approximately 8,502 square kilometers and is an important source of hydroelectric power for the country.

Ghana is known for its rich cultural heritage: Ghana has a rich cultural heritage, with a mix of traditional African and European influences. The country is known for its vibrant music and dance scene, with traditional drumming and dancing being a staple of local festivals and celebrations.

Ghana is a major producer of gold: Ghana is one of the largest producers of gold in Africa, and the industry is a major contributor to the country's economy. The country is also a major producer of other minerals, including diamonds and bauxite.

Ghana is the birthplace of former United Nations Secretary-General Kofi Annan: Kofi Annan, who was born in Ghana in 1938, served as the Secretary-General of the United Nations from 1997 to 2006. He was the first person from sub-Saharan Africa to hold the position and was awarded the Nobel Peace Prize in 2001 for his efforts to promote peace and security in the world.

Ghana is home to several stunning waterfalls: Ghana is home to several beautiful waterfalls, including the Kintampo Falls and the Wli Falls. These waterfalls are popular tourist destinations and are known for their breathtaking beauty and unique features.

Guinea

Guinea is home to the Mount Nimba Strict Nature Reserve: The Mount Nimba Strict Nature Reserve is located in the southeast of the country and is known for its stunning scenery and rich wildlife. The reserve is home to a wide variety of animals, including elephants, monkeys, and several species of primates.

Guinea is the birthplace of Alpha Conde: Alpha Conde, who was born in Guinea in 1938, is a political leader and former opposition activist who has served as the President of Guinea since 2010. He is known for his efforts to promote democracy, human rights, and economic development in the country.

Guinea is known for its rich musical heritage: Guinea is home to a vibrant music scene, with a rich tradition of traditional music and dance. Some of the most popular traditional music styles in Guinea include balafon, djembe, and kora.

Guinea is a major producer of bauxite: Bauxite is a type of rock that is the primary source of aluminum and is used to make a wide variety of products, including aluminum cans and automobiles. Guinea is one of the largest producers of bauxite in the world, and the industry is a major contributor to the country's economy.

Guinea is home to several beautiful waterfalls: Guinea is home to several stunning waterfalls, including the Kinkon Falls and the Sorsorsor Falls. These waterfalls are popular tourist destinations and are known for their breathtaking beauty and unique features.

Guinea-Bissau

Guinea-Bissau is a country located in West Africa, bordered by Senegal to the north and Guinea to the south and east. It has a population of approximately 1.9 million people and its capital is Bissau.

Guinea-Bissau has a rich cultural heritage, with a mix of African, Portuguese, and Brazilian influences. The country is known for its beautiful beaches, mangrove forests, and rich wildlife, including a large number of bird species.

Despite its natural wealth, Guinea-Bissau is one of the poorest countries in the world, with a high poverty rate and a low standard of living. The country has faced numerous political and economic challenges, including coups, civil war, and instability, which have hindered its development.

In recent years, there have been efforts to improve the situation in Guinea-Bissau, including the establishment of a new government and the implementation of economic reforms. Despite these efforts, the country continues to face significant challenges, and much work remains to be done to ensure stability and prosperity for its people.

The country is located at the crossroads of West Africa, and its history and culture have been shaped by the influence of several neighboring countries, including Senegal, Mali, and Sierra Leone.

Kenya

Kenya is home to the Maasai Mara National Reserve: The Maasai Mara National Reserve is one of the most popular tourist destinations in Kenya, known for its incredible wildlife and breathtaking scenery. Visitors to the reserve can see a wide variety of wildlife, including lions, cheetahs, elephants, and giraffes.

Kenya is the birthplace of marathon runner Eliud Kipchoge: Eliud Kipchoge, who was born in Kenya in 1984, is considered one of the greatest marathon runners of all time. He has won numerous marathons, including the Olympic Games, and has set multiple world records.

Kenya is known for its vibrant culture and traditional dances: Kenya has a rich and vibrant culture that is reflected in its music, dances, and traditional costumes. Some of the most popular traditional dances in Kenya include the Kikuyu ngoma, the Kalenjin lelwow, and the Luo ohangla.

Kenya is home to Mount Kenya, the second-highest peak in Africa: Mount Kenya is the second-highest peak in Africa and is an important symbol of national pride for Kenyans. It is also a popular destination for hikers and climbers, who come from all over the world to explore its stunning scenery and challenging trails.

Kenya is a major producer of tea and coffee: Kenya is one of the world's largest producers of tea and coffee, and both of these crops are major contributors to the country's economy. The country's high-quality tea and coffee are known for their unique flavors and are highly sought after by consumers around the world.

Lesotho

Lesotho is a small country located in southern Africa, and it is known for its high elevation, with most of the country sitting at over 1,000 meters above sea level. In fact, the lowest point in Lesotho is still 1,400 meters above sea level, which is higher than the highest point in many countries. This unique geography has earned Lesotho the nickname "Kingdom in the Sky."

Lesotho is known for its unusual tradition of "firewalking," where people walk over hot coals as a demonstration of their bravery and courage.

Lesotho is a landlocked country, meaning that it does not have a coastline. However, what makes Lesotho even more unique is that it is completely surrounded by another country: South Africa. This means that in order to travel to and from Lesotho, you must cross through South African territory. This geographic arrangement is known as an enclave, and Lesotho is one of only a few enclaves in the world.

The Basotho people of Lesotho are known for their unique headgear, the Mokorotlo. The Mokorotlo is a conical hat made from woven grass, and it is an important part of traditional Basotho dress. The hat is so iconic that it has even been featured on the flag of Lesotho. Wearing the Mokorotlo is a symbol of cultural pride for the Basotho people, and it is still a common sight in Lesotho today.

Donkey Derby: Lesotho is home to the annual Donkey Derby, a local tradition and sporting event that takes place in the town of Morija. In the Donkey Derby, participants race their donkeys over a set course, and the winner is crowned the champion donkey racer of the year. The Donkey Derby is a fun and lighthearted event that attracts spectators from all over the country, and it is a great example of the unique culture and traditions of Lesotho.

Liberia

Liberia is the only African country founded by free people of color from the United States, who returned to Africa in the 19th century as part of the American Colonization Society.

The capital of Liberia, Monrovia, is named after James Monroe, the fifth president of the United States.

Liberia is home to many species of primates, including the pygmy chimpanzee, also known as the bonobo. This species is known for its playful and peaceful behavior.

Liberia is one of the few African countries where baseball is played, and it is a popular sport among both children and adults.

The national symbol of Liberia is the ivory-billed woodpecker, a large and striking bird that is native to the country's forests. The bird is so iconic that it is even featured on the Liberian flag.

Libya

Libya is the fourth largest country in Africa: Libya is a large, predominantly desert country located in North Africa. It is the fourth largest country in Africa, after Algeria, Democratic Republic of the Congo, and Sudan.

Libya has a rich history and cultural heritage: Libya has a rich history and cultural heritage, with a mix of Berber, Arab, and African influences. The country is home to several important historical sites, including the ancient city of Leptis Magna and the Cyrene ruins, which are both listed as UNESCO World Heritage Sites.

Libya is a major oil producer: Libya is one of the largest oil producers in Africa and the world, and the industry is a major contributor to the country's economy. The country is home to several large oil fields and has a long history of oil production.

Libya is a large country with a small population: Despite its large size, Libya has a relatively small population. The country's population is estimated to be around 6.8 million people.

Libya has a diverse landscape: Libya has a diverse landscape, ranging from the vast deserts of the Sahara to the lush green hills of the Nafusa Mountains. The country is also home to several lakes and oases, including the Tadrart Acacus, a protected area in the southwestern part of the country that is known for its stunning rock formations and diverse wildlife.

Madagascar

Madagascar is home to many unique species of animals: Madagascar is known for its high level of biodiversity, with many species of plants and animals found nowhere else in the world. Some of the most famous of these include the lemur, a type of primates that are only found on the island, as well as the fossa, a large predator that is similar to a cougar.

Madagascar has a rich cultural heritage: Madagascar has a rich cultural heritage, with a mix of African, Asian, and European influences. The island is home to many different ethnic groups, each with their own unique traditions and customs.

Madagascar is the fourth largest island in the world: Madagascar is the fourth largest island in the world, after Greenland, New Guinea, and Borneo. The island is located off the coast of East Africa and is separated from the mainland by the Mozambique Channel.

Madagascar is known for its beautiful beaches: Madagascar is home to many beautiful beaches, including Nosy Be, a popular tourist destination located off the northwest coast of the island. The beaches of Madagascar are known for their clear blue waters, white sand, and stunning sunsets.

Madagascar is one of the world's leading producers of vanilla: Madagascar is one of the world's leading producers of vanilla, a flavor that is widely used in cooking and baking. The country produces about 80% of the world's supply of vanilla and is one of the largest exporters of the spice.

Malawi

Malawi is a landlocked country in southeast Africa, known for its stunning natural beauty and diverse landscapes, including the stunning Lake Malawi, which is home to hundreds of species of fish and is a popular destination for water sports and beach-goers.

Malawi is home to several unique species of wildlife, including the African Elephant, the African Lion, and the Thornicroft's Giraffe, which is found only in southern Malawi.

Malawi has a rich cultural heritage, with a diverse population of ethnic groups, including the Chewa, Lomwe, and Yao people. This diversity is reflected in the country's traditional music, dance, and festivals, which celebrate the unique cultural traditions of each group.

Malawi is known for its vibrant food scene, with a cuisine that is heavily influenced by the country's agricultural heritage. Popular dishes include nsima, a type of maize porridge, and relish, a type of vegetable stew made with ingredients like peanuts, beans, and corn.

Malawi is a country with a rich history, including the Maravi Kingdom, which dominated the region in the 16th and 17th centuries, and the British colonial period, which lasted from the late 19th century to the mid-20th century. Visitors can explore the country's many historical sites and monuments, including the Chongoni Rock Art Area, which features rock paintings dating back thousands of years.

Mali

Mali is home to Timbuktu, one of the world's most famous historical cities: Timbuktu, located in northern Mali, is one of the world's most famous historical cities and was once a center of trade, scholarship, and Islamic culture. Today, Timbuktu is a UNESCO World Heritage Site and is known for its beautiful architecture and rich cultural heritage.

Mali is one of the world's poorest countries: Despite its rich cultural heritage and history, Mali is one of the poorest countries in the world, with a high level of poverty and limited access to basic services such as healthcare and education.

Mali is home to the Niger River, one of Africa's largest rivers: The Niger River is one of the largest rivers in Africa and is an important source of water for irrigation, fishing, and transportation in Mali and other countries in West Africa.

Mali is known for its rich musical tradition: Mali is known for its rich musical tradition, with a long history of musical performance and production. The country is home to many famous musicians, including Ali Farka Touré and Amadou & Mariam, who have gained international recognition for their musical talent.

Mali was once a part of the powerful Mali Empire: Mali was once a part of the powerful Mali Empire, which was one of the largest and richest empires in West Africa from the 13th to the 16th centuries. The empire was known for its wealth, power, and cultural achievements, and its legacy can still be seen in cultural heritage of modern-day Mali.

Mauritania

Mauritania is a country located in West Africa, and is known for its stunning natural beauty, including the vast deserts of the Sahara, the rugged landscapes of the Adrar Plateau, and the rolling dunes of the Erg Chebbi.

Mauritania is home to several unique and endemic species of wildlife, including the Addax antelope, the Dorcas gazelle, and the Saharan cheetah. These species can be found in the country's many national parks and wildlife reserves.

Mauritania has a rich cultural heritage, with a diverse population of ethnic groups, including the Moors, the Soninke, and the Wolof people. This diversity is reflected in the country's traditional music, dance, and festivals, which celebrate the unique cultural traditions of each group.

Mauritania is known for its rich cuisine, with a food scene that is heavily influenced by the country's nomadic heritage. Popular dishes include thieboudienne, a type of fish and rice dish, and mafé, a type of stew made with peanut butter and vegetables.

Mauritania has a long and rich history, including the ancient Kingdom of Mauretania, which dominated the region from the 3rd century BC to the 3rd century AD, and the medieval Islamic state of Tagharga, which ruled over parts of the country from the 11th to the 16th centuries. Visitors can explore the country's many historical sites and monuments, including the Chinguetti Mosque, one of the oldest mosques in West Africa, and the Ksar Ouled Soltane, a traditional mud-brick city.

Mauritius

Mauritius is home to a unique species of flightless bird called the "Mauritius kestrel". This bird was once on the brink of extinction with only four individuals remaining, but has since made a successful comeback due to conservation efforts.

The island nation is surrounded by crystal-clear waters, making it a popular destination for water sports such as snorkeling and scuba diving. The island is also home to many coral reefs and shipwrecks, providing excellent opportunities to explore underwater wildlife and shipwrecks.

The famous French botanist and explorer, Pierre Poivre, introduced spices such as nutmeg, clove, and cinnamon to the island in the 18th century, turning Mauritius into a major center of spice production.

The capital city of Port Louis is home to the oldest theater in the southern hemisphere, the "Theatre Royal", which was built in 1822.

The "Seven Colored Earths" in Chamarel, Mauritius, is a geological phenomenon where sand dunes are formed from seven different colored sands, including red, brown, green, blue, purple, yellow, and violet. This unique formation is a popular tourist attraction in Mauritius.

Morocco

Morocco is known for its vibrant culture and traditional crafts, including intricate tilework, weaving, and leather goods. The city of Fes is considered the center of traditional Moroccan crafts, with many workshops and artisanal shops.

Morocco has a rich culinary tradition, with dishes that incorporate a mix of spices, herbs, and fresh ingredients. Some of the most famous dishes include tagine, a slow-cooked stew, and couscous, a staple grain dish.

Morocco is home to the world-famous Sahara Desert, one of the largest deserts in the world. Visitors can take camel treks and camp in the desert to experience its stunning beauty and unique landscape.

The city of Marrakesh is famous for its historic medina, a walled old city with narrow winding streets and bustling marketplaces. The city is also home to several historic palaces and gardens, including the famous Bahia Palace.

Morocco has a rich history, with roots in Berber, African, Arab, and European cultures. The country was home to several powerful empires, including the Berber Empire, the Almoravid dynasty, and the Marinid dynasty, and has been influenced by its interactions with these civilizations.

Mozambique

Mozambique is a country located in southeastern Africa, known for its stunning natural beauty, including the sparkling waters of the Indian Ocean, the lush tropical forests of the Gorongosa National Park, and the rolling hills of the Zambesi Basin.

Mozambique is home to a rich and diverse array of wildlife, including several species of primates, big cats, and elephants. Visitors can explore the country's many national parks and wildlife reserves, including the Limpopo National Park and the Banhine National Park.

Mozambique has a rich cultural heritage, with a diverse population of ethnic groups, including the Shona, the Sena, and the Makua people. This diversity is reflected in the country's traditional music, dance, and festivals, which celebrate the unique cultural traditions of each group.

Mozambique is known for its delicious cuisine, with a food scene that is heavily influenced by the country's Portuguese colonial history. Popular dishes include matapa, a type of peanut stew, and prawn peri-peri, a dish made with spicy prawns.

Mozambique has a rich history, including the arrival of Portuguese explorers in the late 15th century and the country's role as a major center of the slave trade in the 16th and 17th centuries. Visitors can explore the country's many historical sites and monuments, including the Fortress of Lourenço Marques and the Maputo Cathedral.

Namibia

Namibia is home to the largest sand dunes in the world, located in the Namib Desert. The dunes can reach up to 1,000 feet tall and provide a unique and stunning landscape for visitors to explore.

Namibia is known for its diverse and unique wildlife, including elephants, lions, cheetahs, and many species of antelopes and birds. It is also home to the Etosha National Park, which is one of the largest game reserves in Africa.

The Himba people, one of Namibia's indigenous ethnic groups, are known for their distinctive red body paint, intricate jewelry, and traditional way of life. They live in remote areas of northern Namibia and maintain many of their ancient customs and beliefs.

The Skeleton Coast, located in northern Namibia, is a remote and rugged coastline known for its shipwrecks and whale bones. The coast is a popular destination for adventurous travelers who want to explore its haunting beauty and rich history.

Namibia was the first country in the world to incorporate environmental protection into its constitution, with a commitment to conserving its wildlife and natural resources. This has made Namibia a leader in African conservation efforts and a popular destination for eco-tourism.

Niger

Niger is the largest country in West Africa and is home to a diverse mix of ethnic groups and cultures. The Hausa and Zarma people are two of the largest ethnic groups in the country, and each have their own distinct language, customs, and traditions.

Niger is known for its vibrant music and dance scenes, with a variety of traditional instruments and styles that reflect its diverse cultural heritage. The tambour drum, the goge flute, and the kora harp are just a few of the instruments used in traditional music and dance performances.

The city of Agadez, located in northern Niger, is known for its unique architecture and cultural heritage. The city is home to the famous Agadez Mosque, which is considered one of the most beautiful examples of West African Islamic architecture.

The desert landscape of Niger is home to several unique wildlife species, including the addax antelope, the Saharan cheetah, and the desert gazelle. The country is also home to several large wildlife reserves and national parks, offering opportunities for wildlife viewing and exploration.

Niger is known for its rich cultural heritage, with a long history of kingdoms and empires that have shaped the region over the centuries. The ancient city of Jenne, located in the southwestern part of the country, is considered one of the oldest cities in Africa and is home to many well-preserved historic buildings and artifacts.

Nigeria

Nigeria is known for its vibrant and diverse music scene, with a wide range of styles that reflect the country's cultural heritage. Styles such as Highlife, Juju, and Fuji music are popular and widely enjoyed both within Nigeria and around the world.

Nigeria has a rich and diverse cultural heritage, with over 500 ethnic groups and over 500 different languages spoken throughout the country. This diversity is reflected in the country's traditional festivals, crafts, and cuisines, which are unique to each region and ethnic group.

The city of Lagos, located in southwestern Nigeria, is the largest city in Africa and is known for its vibrant nightlife, cultural heritage, and economic opportunities. It is a major center of commerce and finance, and is home to several historic sites, museums, and cultural centers.

Nigeria is home to several unique and endangered species of wildlife, including the drills and chimpanzees, both of which are found in the country's tropical forests. The country is also home to several national parks and wildlife reserves, offering opportunities for wildlife viewing and exploration.

Nigeria is one of the largest oil-producing countries in the world and has a strong and growing economy. It is a major player in the global oil and gas industry and is also home to several thriving industries, including telecommunications, manufacturing, and construction.

Rwanda

Rwanda is known as the "Land of a Thousand Hills," due to its rolling hills and mountainous terrain. The country is home to several breathtaking landscapes, including the Virunga Volcanoes, the savannas of Akagera National Park, and the stunning lakes of the Rwandan highlands.

Rwanda is famous for its gorillas, with over half of the world's mountain gorillas residing in the country's Volcanoes National Park. Visitors can take guided tours to visit the gorillas in their natural habitat and observe these fascinating primates up close.

Rwanda is a leading producer of coffee and tea, with several large plantations located throughout the country. Visitors can tour these plantations, learn about the coffee and tea-making process, and sample the local produce.

The city of Kigali, located in central Rwanda, is the country's capital and largest city. It is a modern and vibrant city, with a growing economy and several cultural and historical sites to visit, including the Kigali Genocide Memorial and the National Museum of Rwanda.

Rwanda is known for its efforts to promote sustainability and conservation, with several initiatives aimed at protecting the country's natural resources and wildlife. The country has made significant strides in protecting its gorillas and other wildlife, and has become a leader in African conservation efforts.

São Tomé and Principe

São Tomé and Principe is a small island nation located in the Gulf of Guinea, off the western coast of Africa. The country is known for its stunning natural beauty, including lush tropical forests, stunning beaches, and rolling hills.

São Tomé and Principe is a biodiversity hotspot, with a rich array of plant and animal species, many of which are endemic to the islands. The country is home to several national parks and wildlife reserves, including the Obô Natural Park and the Bom Successo Botanical Garden.

São Tomé and Principe has a rich cultural heritage, with a population of ethnic groups, including the São Toméan, Angolan, and Portuguese people. This diversity is reflected in the country's traditional music, dance, and festivals, which celebrate the unique cultural traditions of each group.

São Tomé and Principe is known for its delicious cuisine, with a food scene that is heavily influenced by the country's Portuguese colonial history. Popular dishes include feijão, a type of bean stew, and bolo do caco, a type of sweet potato cake.

São Tomé and Principe has a rich history, including the arrival of Portuguese explorers in the late 15th century and the country's role as a major center of the slave trade in the 16th and 17th centuries. Visitors can explore the country's many historical sites and monuments, including the São Tomé and Principe National Museum and the Fort of São Sebastião.

Senegal

Senegal is known for its vibrant music and dance scene, with several popular styles including mbalax, which is a fusion of traditional Senegalese music and Western influences. Senegal is also home to the Dakar International Jazz Festival, which draws musicians and fans from around the world.

Senegal is a diverse country, with a mix of different ethnic groups and cultures, including the Wolof, the Serer, and the Lebu people. This diversity is reflected in the country's food, clothing, and cultural traditions, which are unique to each region and ethnic group.

Senegal is known for its stunning beaches and coastal landscapes, including the Dakar Peninsula, the Pink Lakes of Dakar, and the Petite Côte. These areas are popular destinations for sunbathing, swimming, and other beach activities, and offer breathtaking views of the Atlantic Ocean.

The city of Dakar, located on the western coast of Senegal, is the country's capital and largest city. It is a major center of commerce and culture, with a bustling port, several museums, and a vibrant nightlife scene.

Senegal has a rich cultural heritage, with a history of kingdoms and empires that have shaped the region over the centuries. The Gorée Island, located off the coast of Dakar, is a UNESCO World Heritage site and was once a major center of the African slave trade. Visitors can tour the island's historic sites and learn about its rich cultural and historical legacy.

Seychelles

Seychelles is an archipelago of 115 islands located in the Indian Ocean, east of mainland Africa. The country is known for its stunning beaches, crystal-clear waters, and lush tropical forests, making it a popular destination for tourists and beach-goers.

Seychelles is home to several unique and endemic species of flora and fauna, including the Coco de Mer palm tree, which produces the largest seed in the world, and the Seychelles Warbler, a small bird found only on the islands.

The Seychellois people are a mix of African, Asian, and European descent, reflecting the island's rich cultural heritage. This heritage is reflected in the country's food, music, and festivals, which celebrate the diverse cultural traditions of the Seychellois people.

The Seychelles is a leading destination for marine wildlife and eco-tourism, with several large marine parks and reserves that protect the country's diverse marine life, including giant tortoises, dolphins, and numerous species of fish and coral.

The Seychelles was once a major center of the spice trade, and is known for its production of cinnamon, vanilla, and other spices. Visitors can tour the island's spice plantations and learn about the history of the spice trade in Seychelles and its role in shaping the island's economy and cultural heritage.

Sierra Leone

Sierra Leone is known for its stunning natural beauty, including the lush rainforests of the Tiwai Island Wildlife Sanctuary and the gorgeous beaches of River Number Two. The country is also home to several national parks, including the Outamba Kilimi National Park, which is home to elephants, lions, and other wildlife.

Sierra Leone is a culturally rich country, with a diverse population of ethnic groups, including the Temne, Mende, and Limba people. This diversity is reflected in the country's music, dance, and traditional festivals, which celebrate the unique cultural traditions of each group.

Sierra Leone is one of the world's largest producers of diamonds, and the diamond trade has played a major role in shaping the country's economy and political landscape. Visitors can tour diamond mines and learn about the history and impact of the diamond trade in Sierra Leone.

Sierra Leone is known for its vibrant food scene, with a cuisine that is a fusion of West African, European, and Asian influences. Popular dishes include Jollof Rice, a spicy tomato and rice dish, and Cassava Leaf Stew, a flavorful stew made from cassava leaves and spices.

The city of Freetown, located on the western coast of Sierra Leone, is the country's capital and largest city. It is a vibrant and bustling city, with a rich history and a thriving cultural scene, including several museums, art galleries, and musical venues.

Somalia

Somalia is a country located in the Horn of Africa, and is known for its stunning natural beauty, including the beaches along the Indian Ocean and the rugged landscapes of the Hargeisa and Puntland regions.

Somalia is home to several unique and endemic species of wildlife, including the Somali Wild Ass, the Beira antelope, and the Opole gazelle. These species can be found in the country's many national parks and wildlife reserves.

Somalia has a rich cultural heritage, with a long history of poetry, storytelling, and music. The country is known for its traditional music, which features intricate melodies and rhythms, and is performed with instruments like the oud, the kaban, and the calabash.

Somalia has a rich culinary tradition, with a cuisine that is heavily influenced by the country's nomadic heritage. Popular dishes include camel meat, injera, a type of flatbread made from teff flour, and basbousa, a sweet cake made from semolina flour and honey.

The capital of Somalia is Mogadishu, a city located on the Indian Ocean coast. Mogadishu is a vibrant and bustling city, with a rich history and a thriving cultural scene, including several museums, art galleries, and cultural centers. Despite the challenges posed by conflict and insecurity, the city is undergoing a period of renewal and growth, and is poised to become a major center of culture and commerce in the region.

South Africa

South Africa is a country located in the southern tip of Africa, and is known for its stunning natural beauty, including the majestic Table Mountain, the rolling vineyards of the Western Cape, and the rolling hills and savannas of the Great Karoo.

South Africa is home to an incredible diversity of wildlife, including the Big Five (lion, elephant, leopard, buffalo, and rhinoceros) and other species like the cheetah, the hyena, and the hippopotamus. The country is also home to several national parks and wildlife reserves, including the Kruger National Park and the Hluhluwe-iMfolozi Park.

South Africa has a rich cultural heritage, with a diverse population of ethnic groups, including the Zulu, Xhosa, and Boer people. This diversity is reflected in the country's traditional music, dance, and festivals, which celebrate the unique cultural traditions of each group.

South Africa is known for its vibrant food scene, with a cuisine that is heavily influenced by the country's colonial history, as well as its indigenous and African roots. Popular dishes include bobotie, a type of meat pie made with beef or lamb, and bunny chow, a type of curry served in a hollowed-out loaf of bread.

South Africa has a rich history, including the apartheid era, which lasted from 1948 to the early 1990s, and the transition to democracy, which began in the mid-1990s. Visitors can explore the country's many historical sites and monuments, including

Robben Island, where Nelson Mandela was imprisoned for many years, and the Apartheid Museum in Johannesburg.

South Sudan

South Sudan is a landlocked country located in northeastern Africa, known for its stunning natural beauty, including the lush savannas of the Boma National Park, the rolling hills of the Imatong Mountains, and the sparkling waters of Lake Turkana.

South Sudan is home to a rich and diverse array of wildlife, including several species of primates, big cats, and elephants. Visitors can explore the country's many national parks and wildlife reserves, including the Badingilo National Park and the Bandingilo Wildlife Reserve.

South Sudan has a rich cultural heritage, with a diverse population of ethnic groups, including the Dinka, the Nuer, and the Shilluk people. This diversity is reflected in the country's traditional music, dance, and festivals, which celebrate the unique cultural traditions of each group.

South Sudan is known for its delicious cuisine, with a food scene that is heavily influenced by the country's rich cultural heritage. Popular dishes include dura, a type of porridge made from sorghum, and kisra, a type of flatbread.

South Sudan has a rich history, including the arrival of the first humans in the region over 100,000 years ago, the rise of the Kingdom of Kush in the 8th century BCE, and the country's role as a major center of the slave trade in the 19th century. Visitors can explore the country's many historical sites and monuments, including the Maridi Museum and the Palace of the Nubian Kings in Juba.

Sudan

Sudan is a country located in northeastern Africa, known for its stunning natural beauty, including the rolling sand dunes of the Sahara Desert, the lush savannas of the South Sudanese Plains, and the sparkling waters of the Nile River.

Sudan is home to a rich and diverse array of wildlife, including several species of primates, big cats, and elephants. Visitors can explore the country's many national parks and wildlife reserves, including the Dinder National Park and the Sanganeb National Park.

Sudan has a rich cultural heritage, with a diverse population of ethnic groups, including the Nubians, the Arabs, and the Beja people. This diversity is reflected in the country's traditional music, dance, and festivals, which celebrate the unique cultural traditions of each group.

Sudan is known for its delicious cuisine, with a food scene that is heavily influenced by the country's rich cultural heritage. Popular dishes include kisra, a type of flatbread, and foul, a type of dish made with beans.

Sudan has a rich history, including the rise of the Kingdom of Kush in the 8th century BCE, the arrival of the Islamic empire in the 7th century CE, and the country's role as a major center of the slave trade in the 19th century. Visitors can explore the country's many historical sites and monuments, including the Pyramids of Meroe and the Al-Aqsa Mosque in Omdurman.

Tanzania

Tanzania is a country located in East Africa, known for its stunning natural beauty, including the rolling hills of the

Serengeti, the sparkling waters of Lake Victoria, and the majestic peaks of Mount Kilimanjaro, the highest mountain in Africa.

Tanzania is home to a rich and diverse array of wildlife, including several species of primates, big cats, and elephants. Visitors can explore the country's many national parks and wildlife reserves, including the Serengeti National Park and the Ngorongoro Conservation Area.

Tanzania has a rich cultural heritage, with a diverse population of ethnic groups, including the Chaga, the Sukuma, and the Zaramo people. This diversity is reflected in the country's traditional music, dance, and festivals, which celebrate the unique cultural traditions of each group.

Tanzania is known for its delicious cuisine, with a food scene that is heavily influenced by the country's rich cultural heritage. Popular dishes include ugali, a type of maize porridge, and biryani, a type of rice dish.

Tanzania has a rich history, including the rise of the Kingdom of Zimbabwe in the 11th century CE, the arrival of the Portuguese in the 16th century, and the country's role as a major center of the Arab slave trade in the 19th century. Visitors can explore the country's many historical sites and monuments, including the ruins of the Kilwa Kisiwani and the House of Wonders in Stone Town, Zanzibar.

Togo

Togo is a small country located in West Africa, known for its stunning landscapes, including the rolling hills of the Togo Mountains and the sandy beaches of the Atlantic coast.

Togo has a rich cultural heritage, with a diverse population of ethnic groups, including the Ewe, the Kabye, and the Mina people. This diversity is reflected in the country's traditional music, dance, and festivals, which celebrate the unique cultural traditions of each group.

Togo is known for its delicious cuisine, with a food scene that is heavily influenced by the country's rich cultural heritage. Popular dishes include fufu, a type of starchy food made from yam or cassava, and palm nut soup, a type of spicy soup made with palm oil.

Togo is home to several important cultural and historical sites, including the Tamberma Valley, a region known for its traditional mud-brick buildings, and the Aneho Royal Palace, a former royal palace that is now a museum.

Togo has a rich history, including the rise of the Kingdom of Dahomey in the 17th century CE, the arrival of the French in the late 19th century, and the country's role as a major center of the slave trade in the 19th century. Visitors can explore Togo's many historical sites and monuments, including the Palais des Congrès in Lomé, the country's largest city.

Tunisia

Tunisia is a country located in North Africa, known for its stunning landscapes, including the rolling hills of the Atlas Mountains and the sparkling waters of the Mediterranean Sea.

Tunisia has a rich cultural heritage, with a diverse population of ethnic groups, including the Berbers, the Arabs, and the Jews. This diversity is reflected in the country's traditional music, dance, and festivals, which celebrate the unique cultural traditions of each group.

Tunisia is known for its delicious cuisine, with a food scene that is heavily influenced by the country's rich cultural heritage. Popular dishes include brik, a type of pastry filled with egg, tuna, and vegetables, and couscous, a type of grain dish made with semolina.

Tunisia is home to several important cultural and historical sites, including the ancient city of Carthage, a former Phoenician colony that was one of the most important cities of the ancient world, and the Bardo Museum in Tunis, one of the largest museums of Roman mosaics in the world.

Tunisia has a rich history, including the rise of the ancient city-state of Carthage, the arrival of the French in the 19th century, and the country's role as a major center of the Arab-Islamic civilization in the Middle Ages. Visitors can explore Tunisia's many historical sites and monuments, including the Roman ruins of Dougga and the medina of Tunis, a UNESCO World Heritage Site.

Uganda

Uganda is a country located in East Africa, known for its stunning natural beauty, including the rolling hills of the Rwenzori Mountains, the sparkling waters of Lake Victoria, and the lush jungles of the Bwindi Impenetrable Forest.

Uganda is home to a rich and diverse array of wildlife, including several species of primates, big cats, and gorillas. Visitors can explore the country's many national parks and wildlife reserves, including the Queen Elizabeth National Park and the Murchison Falls National Park.

Uganda has a rich cultural heritage, with a diverse population of ethnic groups, including the Baganda, the Basoga, and the Banyoro people. This diversity is reflected in the country's traditional music, dance, and festivals, which celebrate the unique cultural traditions of each group.

Uganda is known for its delicious cuisine, with a food scene that is heavily influenced by the country's rich cultural heritage. Popular dishes include matoke, a type of steamed plantain, and g-nut sauce, a type of peanut sauce.

Uganda has a rich history, including the rise of the Kingdom of Buganda in the 13th century CE, the arrival of the British in the late 19th century, and the country's role as a major center of the Arab slave trade in the 19th century. Visitors can explore the country's many historical sites and monuments, including the Kasubi Tombs and the Kabaka's Palace in Kampala.

Zambia

Zambia is a landlocked country located in southern Africa, known for its stunning landscapes, including the rolling hills of the Zambian Plateau, the stunning waterfalls of the Zambezi River, and the lush forests of the Lower Zambezi.

Zambia has a rich cultural heritage, with a diverse population of ethnic groups, including the Bantu, the Tonga, and the Nsenga people. This diversity is reflected in the country's traditional music, dance, and festivals, which celebrate the unique cultural traditions of each group.

Zambia is known for its delicious cuisine, with a food scene that is heavily influenced by the country's rich cultural heritage. Popular dishes include nshima, a type of starchy food made from maize meal, and relish, a type of spicy sauce made from vegetables and spices.

Zambia is home to several important cultural and historical sites, including the South Luangwa National Park, one of the largest and most biodiverse national parks in Africa, and the Victoria Falls, a stunning waterfall located on the Zambia-Zimbabwe border that is considered one of the Seven Natural Wonders of the World.

Zambia has a rich history, including the arrival of European explorers in the 19th century, the rise of the powerful Kingdom of the Lozi in the 19th century, and the country's role as a major center of the anti-colonial struggle in the 20th century. Visitors can explore Zambia's many historical sites and monuments, including the Mukuni Village, a traditional village located near the Victoria Falls, and the Livingstone Memorial, a monument dedicated to the famous explorer David Livingstone.

Zimbabwe

Zimbabwe is a landlocked country located in southern Africa, known for its stunning landscapes, including the rolling hills of the Zimbabwean Plateau, the stunning waterfalls of the Zambezi River, and the lush forests of the Eastern Highlands.

Zimbabwe has a rich cultural heritage, with a diverse population of ethnic groups, including the Shona, the Ndebele, and the Tonga people. This diversity is reflected in the country's traditional music, dance, and festivals, which celebrate the unique cultural traditions of each group.

Zimbabwe is known for its delicious cuisine, with a food scene that is heavily influenced by the country's rich cultural heritage. Popular dishes include sadza, a type of starchy food made from maize meal, and stew, a type of spicy sauce made from vegetables and meat.

Zimbabwe is home to several important cultural and historical sites, including the Great Zimbabwe Ruins, a complex of ancient ruins that were once the center of a powerful kingdom, and the Hwange National Park, one of the largest and most biodiverse national parks in Africa.

Zimbabwe has a rich history, including the arrival of European explorers in the 19th century, the rise of the powerful Kingdom of Zimbabwe in the 11th century, and the country's role as a major center of the anti-colonial struggle in the 20th century. Visitors can explore Zimbabwe's many historical sites and monuments, including the Matopos Hills, a beautiful area that was once the center of a powerful kingdom, and the Rhodes Memorial, a monument dedicated to the famous British colonialist Cecil Rhodes.

NORTH AMERICA

North America is a continent located in the Northern Hemisphere and is the third largest of the seven continents. It is bordered by the Arctic Ocean to the north, the Atlantic Ocean to the east, the Pacific Ocean to the west, and Central America to the south. North America is home to three countries: the United States, Canada, and Mexico. The United States is the largest country on the continent and is known for its diverse landscape, rich cultural heritage, and strong economy. Canada is the second largest country and is known for its natural beauty, friendly people, and high standard of living. Mexico is the third largest country and is known for its rich history, vibrant culture, and diverse landscapes, including deserts, tropical forests, and ancient ruins. North America is also home to numerous indigenous peoples, with a rich history and cultural heritage.

There are 23 countries in North America, including the three main countries: the United States, Canada, and Mexico. In addition, there are 20 smaller island nations and territories in the Caribbean, including Cuba, Jamaica, Haiti, the Dominican Republic, and Puerto Rico. These small island nations and territories are often grouped together with the larger mainland countries to form the larger region of North America.

Antigua and Barbuda

Antigua and Barbuda is a twin-island country located in the Caribbean Sea, known for its stunning beaches and crystal-clear waters. The country is composed of two main islands, Antigua and Barbuda, as well as several smaller islands.

Antigua and Barbuda is home to a rich cultural heritage, with a diverse population of people of African, British, and Spanish descent. This diversity is reflected in the country's traditional music, dance, and festivals, which celebrate the unique cultural traditions of each group.

Antigua and Barbuda is known for its delicious cuisine, with a food scene that is heavily influenced by the country's rich cultural heritage. Popular dishes include fungee, a type of cornmeal pudding, and ducana, a type of sweet potato pudding.

Antigua and Barbuda is home to several important cultural and historical sites, including the Nelson's Dockyard National Park, a former naval base that is now a popular tourist attraction, and the Antigua and Barbuda Museum, which showcases the country's rich cultural and historical heritage.

Antigua and Barbuda has a rich history, including its role as an important center of the sugar trade in the 17th and 18th centuries, its participation in the American Revolution, and its eventual independence from Britain in 1981. Visitors can explore the country's many historical sites and monuments, including the Antigua and Barbuda Independence Square, a popular gathering place for cultural events and festivals, and the Antigua and Barbuda Memorial, which honors the country's fallen heroes.

The Bahamas

Pigs are the only species that swim in the famous swimming pigs of the Exuma, Bahamas. They have become a popular tourist attraction and can be found in various other parts of the island as well.

The national bird of the Bahamas is the West Indian Flamingo, which is pink in color due to the pigments in the algae and crustaceans they feed on.

Nassau, the capital of the Bahamas, was once a pirate's den and was once known as the pirate capital of the world. The city has a rich pirate history, which is reflected in its architecture, museums, and festivals.

In the Bahamas, it is considered good luck to kiss a dolphin if it approaches you while you are swimming in the ocean.

The famous "Junkanoo" festival is celebrated annually in the Bahamas and is a wild street party filled with music, dancing, and colorful costumes. The festival is held on Boxing Day and New Year's Day and is a must-see for anyone visiting the country during the holiday season.

Barbados

Rum is a popular spirit made from sugar cane, and it is said to have originated in Barbados in the 17th century. Today, rum production is a major industry on the island, and many distilleries offer tours and tastings.

Barbados was once a major producer of sugar, and there are many historic sugar plantations still standing on the island. Some of these plantations have been turned into museums or tourist attractions, offering visitors a glimpse into the island's rich history.

Barbados is surrounded by turquoise waters and pristine sandy beaches, making it a popular tourist destination for sun-seekers and beach lovers. Some of the most famous beaches in Barbados include Carlisle Bay, Bathsheba Beach, and Crane Beach.

Rihanna was born and raised in Barbados, and she has become one of the most successful musicians in the world. She has won numerous awards and is known for her hit songs, including "Umbrella," "Diamonds," and "We Found Love."

Barbados is known for its rich cultural heritage, which is a blend of African, British, and West Indian influences. This is reflected in the island's music, food, and festivals, which are all important parts of Barbadian culture. The island is famous for its annual Crop Over Festival, a celebration of the end of the sugar cane harvest, which features music, food, and dancing.

Belize

Belize has a rich and diverse cultural heritage, with influences from a variety of different cultures including Maya, African, Spanish, and British. This has created a unique blend of cultural traditions and customs that can be seen in the island's music, food, and festivals.

Belize is located on the Caribbean coast of Central America and is home to the largest coral reef in the Western Hemisphere. The Belize Barrier Reef is a popular tourist destination for scuba diving and snorkeling, and it is home to a vast array of marine life, including hundreds of species of fish, coral, and other sea creatures.

Chocolate is one of the world's most popular treats, and it is said to have originated in Belize. The ancient Maya people were the first to cultivate and use cacao, and they believed that the drink was a gift from the gods. Today, chocolate is still made and consumed in Belize, and it is an important part of the country's culture and economy.

Belize is home to numerous ancient Maya ruins, including the famous site of Xunantunich. These ruins are some of the most important archaeological sites in the world, and they offer a fascinating glimpse into the history and culture of the ancient Maya civilization.

Belize is home to many species of wildlife, including numerous species of birds, reptiles, and mammals. The country is a sanctuary for several endangered species, including the jaguar, the scarlet macaw, and the manatee. Belize is also home to numerous national parks and wildlife reserves, which are important habitats for these and other species of wildlife.

Canada

Canada is the second largest country in the world by area: Canada covers a vast area of land, stretching from the Atlantic Ocean in the east to the Pacific Ocean in the west. With a total land area of 9.98 million square kilometers, Canada is the second largest country in the world, after Russia.

Canada is a culturally diverse country, with a population made up of people from all over the world. This diversity is reflected in the country's many cultural festivals and events, as well as its vibrant cuisine, art, and music.

Canadians are known for their friendly and welcoming nature, and the country has a reputation as one of the safest and most peaceful places in the world. This makes Canada a popular destination for tourists and immigrants alike.

Canada is home to some of the world's most stunning natural landscapes: From the rugged coastline of the Atlantic provinces to the soaring peaks of the Rocky Mountains, Canada is home to some of the world's most stunning natural landscapes. The country is also home to numerous national parks and wildlife reserves, which are important habitats for many species of wildlife.

Canada has a strong and stable economy, with a well-developed infrastructure and a highly educated workforce. The country is a leader in industries such as technology, natural resources, and manufacturing, and it is known for its high standards of living and quality of life.

Costa Rica

Costa Rica is a leader in sustainability: Costa Rica is known for its commitment to sustainability, and the country has taken many steps to reduce its carbon footprint and protect its environment. For example, Costa Rica generates more than 90% of its electricity from renewable sources, and it has set a goal to be carbon neutral by 2021.

Costa Rica is home to incredible biodiversity: Costa Rica is located in the heart of Central America, and it is home to a rich array of plant and animal species. The country is home to numerous national parks and wildlife reserves, which protect some of the most important habitats for many species of wildlife, including monkeys, toucans, and jaguars.

The country is known for its stunning beaches: Costa Rica is located on the Pacific and Caribbean coasts of Central America, and it is home to many beautiful beaches. Some of the most famous beaches in Costa Rica include Manuel Antonio, Santa Teresa, and Tamarindo.

Costa Rica is a popular destination for adventure travelers, and it offers a wide range of activities for visitors, including surfing, kayaking, and zip-lining. The country is also home to several active volcanoes, including Arenal Volcano, which is one of the most active volcanoes in the world.

Costa Rica is a peaceful and stable country, and it is known for its high standards of living and quality of life. The country has a strong education system, and it is a leader in many industries, including tourism, agriculture, and technology. Costa Rica is also a democracy, and it is considered one of the most progressive and forward-thinking countries in the world.

Cuba

Cuba is a country with a rich cultural heritage, and its capital city of Havana is known for its colonial-style architecture and vibrant music scene. The country is famous for its traditional dance styles, such as the salsa and the rumba, and its music is an important part of its cultural identity.

Cuba has a unique political history, and it was one of the first countries in the world to adopt a communist government after the revolution in 1959. The country has been led by the same government for more than 60 years, and it has a reputation for being one of the last remaining socialist states in the world.

Cuba is famous for its cigars, and the country is considered to be the birthplace of the cigar industry. Cuban cigars are considered to be some of the best in the world, and they are renowned for their rich flavor and high quality.

Cuba is home to several beautiful colonial cities, such as Trinidad, Cienfuegos, and Santiago de Cuba. These cities are known for their stunning colonial-style architecture, and they are popular tourist destinations for visitors who are interested in history and culture.

Cuba is a popular destination for beachgoers, and it is home to many beautiful beaches along its coastline. Some of the most famous beaches in Cuba include Varadero, Guardalavaca, and Cayo Santa Maria, and they are popular destinations for sunbathing, swimming, and water sports.

Dominica

Dominica is known as the "Nature Island of the Caribbean": Dominica is a small island nation located in the Caribbean, and it is known for its lush and verdant landscapes. The country is home to several national parks and wildlife reserves, and it is a popular destination for ecotourism and adventure travel.

Dominica is a volcanic island, and it is home to several active and dormant volcanoes. The island is known for its hot springs, geysers, and bubbling mud pools, and it is a popular destination for tourists who are interested in geology and natural history.

Dominica is home to a rich cultural heritage, and it is a melting pot of African, Caribbean, and European influences. The country is known for its music, dance, and traditional festivals, and it is a popular destination for visitors who are interested in culture and history.

Dominica is a leader in sustainability, and it has taken many steps to reduce its carbon footprint and protect its environment. For example, the country generates more than 50% of its electricity from renewable sources, and it has set a goal to become the first carbon-neutral country in the world.

Dominica is a popular destination for adventure travelers, and it offers a wide range of activities for visitors, including hiking, diving, and snorkeling. The country is also home to many scenic waterfalls, such as Trafalgar Falls and Middleham Falls, which are popular tourist destinations.

The Dominican Republic

The Dominican Republic is home to Santo Domingo, which is the oldest city in the New World. Santo Domingo was founded in 1496, and it is known for its rich history and colonial-style architecture.

The Dominican Republic is a popular tourist destination, and it is known for its beautiful beaches, crystal-clear waters, and year-round warm weather. The country is a popular destination for travelers who are looking for sun, sand, and sea.

The Dominican Republic is known for its rich cultural heritage, and it is a melting pot of African, Spanish, and Taíno influences. The country is famous for its music, dance, and traditional festivals, and it is a popular destination for visitors who are interested in culture and history.

The Dominican Republic is a leader in renewable energy, and it has taken many steps to reduce its carbon footprint and protect its environment. For example, the country generates more than 20% of its electricity from renewable sources, and it has set a goal to become a carbon-neutral country by 2050.

The Dominican Republic is known for its rich biodiversity, and it is home to many species of plants and animals that are found nowhere else in the world. The country is a popular destination for eco-tourism, and it offers many opportunities for visitors to explore its natural beauty, such as hiking, birdwatching, and wildlife viewing.

El Salvador

El Salvador is the smallest country in Central America, and it is located on the Pacific coast of the region. Despite its small size, the country is known for its rich culture, history, and natural beauty.

El Salvador is a volcanic country, and it is home to several active and dormant volcanoes. The country is known for its hot springs, geysers, and bubbling mud pools, and it is a popular destination for tourists who are interested in geology and natural history.

El Salvador is a country with a rich cultural heritage, and it is a melting pot of indigenous, Spanish, and African influences. The country is known for its traditional festivals, music, and dance, and it is a popular destination for visitors who are interested in culture and history.

El Salvador is a leader in renewable energy, and it has taken many steps to reduce its carbon footprint and protect its environment. For example, the country generates more than 25% of its electricity from renewable sources, and it has set a goal to become a carbon-neutral country by 2050.

El Salvador is known for its surf culture, and it is a popular destination for surfers from around the world. The country is home to many world-class surf spots, such as Punta Roca, La Libertad, and Sunzal, and it offers many opportunities for visitors to enjoy the waves and catch a few rides.

Grenada

Grenada is known as the "Island of Spice" because it is one of the largest producers of spices in the world. The country is famous for its nutmeg, mace, cloves, cinnamon, and ginger, and it is a popular destination for tourists who are interested in agriculture and food.

Grenada has a rich cultural heritage, and it is a melting pot of African, French, and British influences. The country is known for its traditional festivals, music, and dance, and it is a popular destination for visitors who are interested in culture and history.

Grenada is a tropical paradise, and it is known for its beautiful beaches, crystal-clear waters, and lush vegetation. The country is a popular destination for travelers who are looking for sun, sand, and sea, and it offers many opportunities for visitors to enjoy its natural beauty, such as hiking, snorkeling, and wildlife viewing.

Grenada is home to the world's first underwater sculpture park, which is located off the coast of St. George's. The park features more than 65 sculptures that were created by artist Jason deCaires Taylor, and it is a popular destination for divers and snorkelers.

Grenada is a leader in sustainable tourism, and it has taken many steps to reduce its carbon footprint and protect its environment. For example, the country has adopted a number of eco-friendly practices, such as using renewable energy, promoting responsible waste management, and conserving its natural resources.

Guatemala

Guatemala is home to some of the most spectacular Mayan ruins: Guatemala is home to some of the most spectacular Mayan ruins in the world, including Tikal, Uaxactún, and Yaxhá. The ruins offer a glimpse into the ancient Mayan civilization, and they are a popular destination for tourists who are interested in history and archaeology.

Guatemala is known for its vibrant culture, and it is a country with a rich artistic tradition. The country is famous for its textiles, music, and dance, and it is a popular destination for visitors who are interested in experiencing the local culture.

Guatemala is a bird-watcher's paradise, and it is home to more than 700 species of birds, including toucans, parrots, and quetzals. The country is a popular destination for birders, and it offers many opportunities for visitors to enjoy its avian wildlife, such as hiking, bird-watching tours, and wildlife viewing.

Guatemala is a coffee-growing country, and it is known for its high-quality Arabica coffee beans. The country is one of the largest coffee producers in the world, and it is a popular destination for coffee lovers who are interested in learning more about the coffee-making process.

Guatemala is a leader in sustainable tourism, and it has taken many steps to reduce its carbon footprint and protect its environment. For example, the country has adopted a number of eco-friendly practices, such as using renewable energy, promoting responsible waste management, and conserving its natural resources. This makes it a great destination for travelers who are looking to have a positive impact on the environment while they explore and experience the country.

Haiti

Haiti made history as the first country in the world to be founded by former slaves. The Haitian Revolution, which took place from 1791 to 1804, was the first successful slave revolt in the world and resulted in Haiti gaining its independence from France. This was a significant event in world history and it had a major impact on the abolition of slavery throughout the world.

Haiti was a center of the slave trade and played a key role in the abolition of slavery in the 19th century: During the colonial period, Haiti was one of the largest slave-trading centers in the Caribbean. The slave trade had a significant impact on the country and its people, and it played a crucial role in the abolition of slavery in the 19th century. Haiti was one of the first countries to abolish slavery and the country remains an important symbol of resistance against slavery and oppression.

The country is known for its rich culture, including music, dance and visual arts: Haiti has a rich cultural heritage, and it is a country with a long and proud artistic tradition. The country is famous for its music, dance, and visual arts, and it is a popular destination for visitors who are interested in experiencing the local culture. Haiti has produced many talented artists, including painters, musicians, and writers, who have helped to preserve and promote the country's rich cultural heritage.

Haiti is the birthplace of voodoo, a religion that blends African spiritual traditions with Catholicism: Haiti is the birthplace of voodoo, a religion that blends African spiritual traditions with Catholicism. Voodoo is an important part of Haiti's cultural heritage, and it is still widely practiced in the country today. Voodoo is known for its rituals, which include drumming, dancing, and spirit possession, and it is an important aspect of Haiti's cultural identity.

Haiti is home to stunning natural beauty, including lush forests, pristine beaches, and stunning mountains: Haiti is a country with a rich natural heritage, and it is home to some of the most stunning landscapes in the Caribbean. The country is known for its lush forests, pristine beaches, and stunning mountains, and it is a popular destination for nature lovers who are interested in exploring the country's natural beauty. Haiti is also home to a number of protected areas, including national parks, reserves, and wildlife sanctuaries, which help to preserve the country's natural heritage and promote sustainable tourism.

Honduras

Honduras is home to the world's second largest coral reef: The Mesoamerican Barrier Reef System, which is located off the coast of Honduras, is the world's second largest coral reef. This stunning underwater environment is home to a rich diversity of marine life, including a wide range of fish, coral, and other sea creatures.

Honduras is home to a number of ancient ruins, including the ancient city of Copán, which was once one of the most important cities in the Mayan civilization. The ruins at Copán are considered to be some of the most well-preserved and impressive Mayan ruins in the world.

Honduras is one of the largest coffee producers in the world, and it is known for producing high-quality coffee beans. The country's coffee industry is an important part of the economy, and it provides employment and income for many people in Honduras.

Honduras has a rich cultural heritage, and it is a country with a long and proud history. The country is home to a number of different indigenous groups, each with their own unique culture and traditions, and it is a popular destination for visitors who are interested in experiencing the local culture.

Honduras is a country with a rich natural heritage, and it is home to some of the most stunning landscapes in Central America. The country is known for its lush forests, stunning mountains, and pristine beaches, and it is a popular destination for nature lovers who are interested in exploring the country's natural beauty. Honduras is also home to a number of protected areas, including national parks and wildlife reserves, which help to preserve the country's natural heritage and promote sustainable tourism.

Jamaica

Jamaica is famous for its musical heritage, and it is considered to be the birthplace of reggae music. The island is home to a number of music festivals throughout the year, including Reggae Sumfest, which is considered to be one of the largest reggae festivals in the world.

Jamaica is a world-renowned tourist destination, and it is known for its stunning beaches, lush tropical forests, and clear turquoise waters. The island is a popular destination for travelers who are looking for sun, sand, and sea, and it is also home to a number of historic sites, cultural attractions, and natural wonders that are well worth exploring.

The Blue Mountains are a range of mountains that are located in eastern Jamaica, and they are considered to be one of the most stunning natural attractions on the island. The Blue Mountains are known for their beautiful scenery, including lush forests and rolling hills, and they are a popular destination for hikers, nature lovers, and coffee aficionados.

Jamaica is famous for its delicious cuisine, which is a unique blend of African, Caribbean, and European flavors. The island is famous for its spicy dishes, including jerk chicken and curry goat, and it is also known for its sweet treats, including rum cakes and sweet potato pudding.

Jamaica is a leader in sustainable tourism, and it is committed to preserving its natural heritage and promoting sustainable development. The island is home to a number of protected areas, including national parks and wildlife reserves, and it is working to promote sustainable tourism practices that help to conserve its natural resources and protect its environment. Jamaica is also working to promote sustainable practices throughout its tourism industry, including the use of renewable energy and eco-friendly accommodations, to ensure that visitors can enjoy its beauty and culture for generations to come.

Mexico

Mexico is the third largest country in Latin America by land area, covering over 1.9 million square kilometers. The country is located in North America and borders the United States to the north and Belize and Guatemala to the south.

Mexico has a rich history and cultural heritage, and it is known for its ancient civilizations, including the Aztecs and the Maya. The country is home to a number of historic sites, including the ancient city of Chichen Itza, which is one of the most well-preserved and impressive Mayan ruins in the world.

Mexico is one of the largest oil-producing countries in the world, and it is a major player in the global energy market. The country has a rich history of oil production, and it is home to a number of large oil fields, including the Cantarell Field, which is one of the largest oil fields in the world.

Mexico is famous for its cuisine, which is a delicious blend of indigenous and Spanish flavors. The country is famous for its spicy dishes, including tacos and enchiladas, and it is also known for its sweet treats, including churros and flan. Mexican cuisine is considered to be one of the most diverse and flavorful in the world.

Mexico is a popular tourist destination, and it is known for its stunning beaches, rich cultural heritage, and vibrant cities. The country is a popular destination for travelers who are looking for sun, sand, and sea, and it is also home to a number of historic sites, cultural attractions, and natural wonders that are well worth exploring. Mexico is also known for its rich cultural traditions, including its festivals and traditions, which are celebrated throughout the year.

Nicaragua

Nicaragua is home to the largest fresh water island in the world, known as Ometepe Island. The island is formed by two large volcanoes that are connected by a narrow strip of land and is a popular tourist destination.

Nicaragua is known for its strong and delicious coffee, and it is considered to be one of the best coffee-producing countries in the world. The country's coffee beans are known for their rich, bold flavor and are highly sought after by coffee lovers around the world.

Nicaragua is home to a giant statue of a mosquito, which is located in Managua, the country's capital. The statue is over 50 feet tall and is a popular tourist attraction, with visitors coming from all over the world to take photos with the giant insect.

Nicaragua is known for its unique form of wrestling called "Lucha Libre", which is similar to Mexican wrestling but with its own unique twists and turns. The matches are fast-paced and energetic, and they often feature colorful masks and over-the-top costumes.

Nicaragua is home to a strange and unique attraction called the "Graveyard of the Anchors", which is located on the shores of Lake Nicaragua. The graveyard is filled with old and rusty anchors that were used by ships that once plied the lake's waters. Visitors can explore the graveyard and see the old anchors up close, making it a quirky and unusual tourist destination.

Panama

Panama is most famous for the Panama Canal, which is a man-made waterway that connects the Atlantic Ocean to the Pacific Ocean. The canal was completed in 1914 and it is considered to be one of the engineering marvels of the modern world. It is still used today as a major shipping route, and millions of tons of cargo pass through the canal each year.

Panama is home to an incredible array of plant and animal species, and it is considered to be one of the most biodiverse countries in the world. The country has a number of national parks and wildlife reserves, including the Panama Canal Watershed, which is a protected area that encompasses the entire canal zone.

Panama has a rich and vibrant Afro-Caribbean culture, which is a blend of African, indigenous, and Spanish influences. The country is famous for its music, dance, and cuisine, and it is a popular destination for visitors who are interested in learning about this unique and colorful culture.

Panama has a rich history, and it has been inhabited by indigenous people for thousands of years. The country was a major hub of activity during the Spanish conquest of the Americas, and it played an important role in the California Gold Rush in the mid-19th century. Today, visitors can explore a number of historic sites and monuments that commemorate Panama's rich cultural heritage.

Panama is considered to be a tax haven, which means that it has a low tax rate and favorable tax laws. As a result, the country is home to a number of international businesses and corporations, and it is a popular destination for people looking to set up offshore accounts and minimize their tax liabilities. Despite its reputation as a tax haven, Panama is still considered to be a relatively stable and safe country, with a well-developed infrastructure and a strong economy.

St. Kitts and Nevis

St. Kitts and Nevis is the smallest country in the Americas in terms of both land area and population. The country is made up of two islands, St. Kitts and Nevis, which are located in the Caribbean Sea. Despite its small size, St. Kitts and Nevis has a rich and diverse culture, as well as a thriving tourism industry.

St. Kitts and Nevis has a long history of sugar production, and sugar was once the main industry in the country. The island of St. Kitts was a major center of sugar production during the colonial era, and many of the island's historic buildings and monuments are still visible today, including the former sugar plantations and the sugar mills.

St. Kitts and Nevis is a popular destination for eco-tourism, as the country is home to a number of pristine beaches, lush rainforests, and diverse wildlife. The island of Nevis is particularly well-known for its natural beauty, and visitors can explore the island's lush green hills and crystal-clear waters.

St. Kitts and Nevis is a member of the Commonwealth of Nations, which is an organization of countries that were once part of the British Empire. The country is also a member of the Caribbean Community (CARICOM), and it is a signatory to the United Nations and other international organizations.

St. Kitts and Nevis has a rich cultural heritage, and the country is home to a number of historic monuments and landmarks, including the Brimstone Hill Fortress National Park, which is a UNESCO World Heritage Site. The country is also famous for its music, dance, and cuisine, which are all heavily influenced by the country's African, European, and indigenous cultural heritage.

St. Lucia

St. Lucia is a small island nation located in the Caribbean, and it is known for its stunning natural beauty, including lush rainforests, pristine beaches, and towering volcanic mountains. The island's most famous landmark is the Pitons, two volcanic mountains that rise dramatically from the sea.

St. Lucia was a British colony for over 200 years, and as a result, the island has a rich cultural heritage that is heavily influenced by both Africa and Europe. St. Lucia is a member of the Commonwealth of Nations, and English is widely spoken on the island.

St. Lucia is home to the St. Lucia Jazz & Arts Festival, which is one of the largest and most prestigious jazz festivals in the Caribbean. The festival attracts some of the biggest names in jazz and other musical genres, and it is a major draw for tourists from around the world.

St. Lucia is a popular destination for adventure sports, including hiking, snorkeling, and scuba diving. The island is home to a number of pristine beaches and crystal-clear waters, making it a popular destination for water sports enthusiasts.

St. Lucia is home to a number of unique wildlife species, including the St. Lucia parrot, which is the country's national bird. The island is also home to a number of other unique species, including the St. Lucia racer snake, the St. Lucia shy anole, and the St. Lucia lappet moth.

St. Vincent and the Grenadines

The country of St. Vincent and the Grenadines is a unique and fascinating place, with a rich culture and history that is reflected in its people and traditions. Made up of 32 islands and cays, only 9 of them are inhabited, making it a perfect destination for those seeking a serene and peaceful getaway.

One of the most exciting events that takes place in St. Vincent and the Grenadines is the Carnival celebrations, which are held during the summer months. These celebrations are known for their vibrant and colorful atmosphere, with music, dancing, and street parades that attract visitors from all over the world.

Another notable aspect of St. Vincent and the Grenadines is its musical heritage. The country is considered the birthplace of calypso music, a genre that originated in the Caribbean in the 20th century and has since spread to other parts of the world. Calypso music is characterized by its upbeat rhythms, catchy melodies, and political and social commentary.

St. Vincent and the Grenadines is also home to the world's only drive-in volcano, the Soufrière Volcano. This unique geological feature last erupted in 1979, and visitors can drive right up to the crater and look inside, making it a truly remarkable and once-in-a-lifetime experience.

Vincent and the Grenadines is a place of great natural beauty, with miles of pristine beaches and lush tropical forests. The country is one of the few in the world where sea turtles come to lay their eggs on the shores each year, making it a haven for wildlife lovers and nature enthusiasts. The hawksbill and leatherback turtles can be seen nesting on the shores from March to September, providing a rare and captivating spectacle for visitors.

Trinidad and Tobago

Trinidad and Tobago is a unique and culturally rich country located in the southern Caribbean. It is known for its diverse population and vibrant mix of African, Indian, Chinese, and European cultures.

Trinidad and Tobago is famous for its Carnival celebrations, which are held each year just before Lent. This event is known for its vibrant and colorful costumes, music, and dancing, and attracts visitors from all over the world.

The country is also known for its rich musical heritage, particularly in the genres of calypso and soca. These styles of music have their roots in Trinidad and Tobago and are now popular around the world, known for their upbeat rhythms and catchy melodies.

Trinidad and Tobago is home to the Asa Wright Nature Centre, a protected wildlife reserve that is considered one of the best birdwatching destinations in the world. The reserve is home to over 170 species of birds, as well as monkeys, armadillos, and other wildlife.

Trinidad and Tobago is famous for its delicious cuisine, which reflects the country's diverse cultural heritage. From roti and doubles to curry and chow, the country's food is a delicious melting pot of flavors and ingredients. Additionally, the country is known for its production of premium rum, with several distilleries located on the islands.

United States of America

The United States of America is a large and culturally diverse country with a rich history and unique traditions. It is made up of 50 states and is the third largest country in the world by land area.

The country is famous for its Independence Day celebrations, which take place on July 4th each year. This event is marked by parades, fireworks, and patriotic gatherings, and is a time for Americans to come together and celebrate their nation's history and heritage.

The United States is also known for its diverse cuisine, which reflects the country's melting pot of cultures and traditions. From classic American comfort foods like hamburgers and hot dogs to regional specialties like Tex-Mex and Cajun, the country has something to offer for every palate.

The country is home to many iconic landmarks and natural wonders, including the Grand Canyon, Niagara Falls, and the Redwoods National Park. These breathtaking destinations attract millions of visitors each year and are considered some of the most beautiful places on earth.

The United States is also renowned for its contributions to popular culture, including movies, music, and television. Hollywood, located in Los Angeles, is considered the entertainment capital of the world, and the country's music industry is famous for producing some of the biggest names in rock, pop, and hip-hop. Additionally, the country is home to many renowned universities and research institutions, making it a hub for innovation and scientific advancement.

SOUTH AMERICA

South America is the fourth largest continent in the world and is located in the western hemisphere. It is bordered by the Atlantic Ocean to the east and the Pacific Ocean to the west. The continent is home to a diverse range of cultures, languages, and landscapes, including vast forests, towering mountain ranges, and sprawling deserts. South America is home to numerous indigenous cultures, as well as a mix of colonial, African, and European influence, making it a unique and fascinating place to explore. The continent is home to a number of iconic cities, including Rio de Janeiro, Buenos Aires, and Bogotá, as well as some of the world's most breathtaking natural wonders, including the Amazon rainforest, the Andes mountains, and the Iguazu Falls.

There are 12 recognized sovereign countries in South America. Some territories, such as French Guiana, are considered overseas departments of other countries and are not considered independent nations. Additionally, some territories like the Falkland Islands, South Georgia, and South Sandwich Islands are also considered part of South America but are territories of other countries, such as the United Kingdom. These territories, while part of the continent, do not have the same international recognition as independent countries.

Argentina

Argentina is a large and culturally rich country located in South America, known for its stunning natural beauty and rich history. It is the eighth largest country in the world by land area and is home to over 44 million people.

One of the most iconic cultural events in Argentina is the Tango, a dance that originated in Buenos Aires in the late 19th century. Today, the Tango is an important part of Argentine culture, with dance schools and performance venues found throughout the country.

Argentina is also known for its delicious cuisine, which is heavily influenced by Spanish and Italian traditions. Some of the most famous dishes include asado (barbecue), empanadas (stuffed pastries), and dulce de leche (a sweet caramel sauce).

The country is home to many stunning natural wonders, including the Iguazu Falls, a series of waterfalls located near the border with Brazil. These falls are considered one of the most beautiful waterfalls in the world and attract millions of visitors each year.

Argentina is also renowned for its rich history, particularly its association with the Argentine revolutionary, José de San Martín, who played a key role in the country's fight for independence from Spain in the early 19th century. Additionally, Argentina is home to many important historical and cultural sites, including the famous Recoleta Cemetery in Buenos Aires, which is the final resting place of many famous Argentine figures, including Eva Perón.

Bolivia

Bolivia is the only country in South America to have a national football (soccer) team made entirely of indigenous players.

Bolivia is famous for its traditional indigenous festivals, which are held throughout the year and reflect the country's diverse cultural heritage. These festivals are celebrated with music, dance, and colorful costumes, and are an important part of Bolivian cultural identity.

The country is also known for its stunning natural wonders, including the Salar de Uyuni, the world's largest salt flat, and Lake Titicaca, which is the highest navigable lake in the world. These destinations are popular among tourists and are considered some of the most unique and beautiful places in the world.

Bolivia is rich in history and is home to many important cultural and historical sites, including the Tiwanaku ruins, an ancient city that was once the center of a powerful civilization. Additionally, Bolivia was an important player in the Latin American Wars of Independence and is home to many monuments and museums that commemorate this period of its history.

Bolivia is known for its delicious cuisine, which is heavily influenced by indigenous, Spanish, and Andean traditions. Some of the most famous dishes include pique a lo macho (a spicy beef and sausage dish), salteñas (a type of empanada), and chicha (a traditional corn-based drink). These dishes are enjoyed by Bolivians and visitors alike, and are an important part of the country's culinary heritage.

Brazil

The Cristo Redentor (Christ the Redeemer) statue in Rio de Janeiro is one of the most iconic landmarks in Brazil and is considered one of the Seven Wonders of the Modern World. The statue stands 30 meters tall and offers panoramic views of the city and the surrounding bay.

The country is famous for its diverse and vibrant music scene, including styles like samba, bossa nova, and forró. These musical styles are an important part of Brazilian cultural identity and are enjoyed by people all over the world.

Brazil is home to the largest carnival celebration in the world, held annually in Rio de Janeiro. The festival attracts millions of visitors from all over the world and is known for its elaborate costumes, music, and dance.

The country is also known for its diverse and flavorful cuisine, which is heavily influenced by indigenous, African, and Portuguese traditions. Some of the most famous dishes include feijoada, a hearty stew made with black beans and meat, and açai, a type of berry used in smoothies and other dishes.

Brazil is home to the largest number of threatened species in the world, including the Amazonian manatee, the jaguar, and the giant otter. The country is working to protect its biodiversity and conserve its unique and important ecosystems.

Chile

Chile is a long and narrow country located in South America, stretching over 4,300 km along the western coast of the continent. Despite its small land area, Chile is incredibly diverse, with a range of natural wonders, cultural attractions, and modern cities.

The country is known for its stunning natural beauty, including the Atacama Desert, one of the driest deserts in the world, and the Torres del Paine National Park, a rugged landscape of glaciers, mountains, and turquoise lakes.

Chile is also famous for its wine, particularly its delicious red wines. The country is home to numerous vineyards and wineries, and is considered one of the top wine-producing regions in the world.

Chile is home to many important cultural attractions, including the Chiloé Archipelago, a group of islands with a rich history and cultural heritage, and the cities of Valparaiso and Santiago, which are known for their vibrant street art scenes and colonial architecture.

Chile is known for its innovative cuisine, which combines traditional indigenous ingredients with modern culinary techniques. Some of the most famous dishes include curanto (a type of stew), pastel de choclo (a type of corn pie), and empanadas (a type of turnover filled with meat, cheese, or vegetables). These dishes are enjoyed by Chileans and visitors alike and are an important part of the country's culinary heritage.

Colombia

Colombia is home to a wide variety of ecosystems, including tropical forests, deserts, and coastal areas, which has resulted in a rich diversity of plants and animals. There are over 56,000 species of plants and animals in Colombia, making it the second most biodiverse country in the world after Brazil.

Colombia is the only South American country with coasts on both the Pacific Ocean and the Caribbean Sea, which provides the country with a diverse range of coastal landscapes, including beaches, mangroves, and coral reefs.

Colombia is the world's largest producer of emeralds, which are highly valued for their rich green color. The country's high-quality emeralds are found in mines located in the Andes Mountains, and have been mined for over 500 years.

The national symbol of Colombia is the Andean Condor, which is one of the largest birds in the world with a wingspan of up to 10 feet. These majestic birds are native to the Andes Mountains and are known for their incredible soaring abilities.

Shakira, one of the world's most popular Latin American pop stars, was born and raised in Colombia. She has sold over 75 million records worldwide and is known for her unique blend of Latin, Arabic, and rock music.

Ecuador

Ecuador's location on the equator means that it experiences year-round warm temperatures and is one of only a few countries in the world where you can stand with one foot in the Northern Hemisphere and one foot in the Southern Hemisphere.

Ecuador is a popular destination for eco-tourism, with a variety of unique ecosystems, including the Amazon rainforest, the Galapagos Islands, and the Andes Mountains. These areas offer visitors the opportunity to see a diverse range of wildlife, including rare species of plants and animals, as well as scenic landscapes and cultural experiences.

The Galapagos Islands are a unique and famous destination, known for their diverse wildlife and geological features. The islands are considered a natural laboratory for the study of evolution, as Charles Darwin's observations there helped form the basis of his theory of evolution.

Mount Chimborazo is the highest active volcano in the world, and is considered a challenging climb for mountaineers. Despite being smaller in height than other famous peaks like Mount Everest, its location near the equator means that its peak is the farthest point from the center of the Earth.

Ecuador is known for its traditional textiles and handmade crafts, particularly those made by indigenous communities in the Andes Mountains. These textiles and crafts are known for their bright colors and intricate designs, and are highly sought after by tourists and collectors.

Guyana

Guyana is located in South America and is bordered by Venezuela, Brazil, and Suriname. It is the only English-speaking country in South America, with English being the official language.

Guyana is known for its rich biodiversity, with over 80% of the country being covered by lush tropical rainforests. These forests are home to a wide variety of plants and animals, including rare and endangered species.

Kaieteur Falls is one of Guyana's most famous natural wonders, and is considered one of the largest single-drop waterfalls in the world. The falls are located in the Kaieteur National Park, which is a protected area known for its scenic beauty and rich wildlife.

The Guyana Shield is a large geological formation that covers a large part of Guyana and includes the ancient Roraima plateau. The Guyana Shield is known for its unique rock formations, including tepuis (table-top mountains), which are famous for their stunning landscapes and unique plant and animal life.

Guyana has a rich history of sugar production, and sugarcane remains one of the country's main agricultural products. Sugar was introduced to Guyana in the 17th century and has been a major part of the country's economy for hundreds of years.

Paraguay

Paraguay is a landlocked country located in South America, bordered by Argentina, Brazil, and Bolivia. It is one of the smallest countries in South America in terms of both land area and population.

Paraguay is home to a variety of stunning natural landscapes, including the Paraná River, the largest river in South America after the Amazon, and the Iguazú Falls, one of the largest waterfall systems in the world.

Paraguay has a rich cultural heritage, with influences from indigenous, Spanish, and German cultures. This is reflected in the country's traditional music, dance, and architecture. The Paraguayan harp is a famous traditional instrument, and Paraguay is known for its distinctive style of traditional music called "polka."

Paraguay is one of the few countries in the world where an indigenous language, Guaraní, is widely spoken and officially recognized as one of the country's official languages alongside Spanish. Guaraní is a vibrant and rich language, with its own literature, music, and cultural traditions.

Paraguay has a rich political history, with a long tradition of political and military leaders who have shaped the country's development. Despite its small size, Paraguay has played a significant role in regional affairs, including as a founding member of the Mercosur trade bloc.

Peru

Machu Picchu, the ancient Inca city located in the Andes Mountains, is one of the most popular tourist destinations in Peru and is considered a wonder of the world.

Lake Titicaca, located between Peru and Bolivia, is the highest navigable lake in the world and is considered a sacred site by the local indigenous people.

The Nazca Lines, a series of geoglyphs located in the Nazca Desert, are one of the world's greatest mysteries. Some believe they were created for astronomical purposes, while others believe they were made as part of religious ceremonies.

The Peruvian national symbol, the vicuña, is a type of South American camelid that is highly prized for its fine wool. The vicuña was almost hunted to extinction, but conservation efforts have helped to protect the species.

The War of the Pacific, a war fought between Peru and Bolivia on one side and Chile on the other, resulted in the loss of Bolivia's access to the sea and is still a source of tension between the countries.

Suriname

Suriname is the smallest country in South America in terms of both land area and population. Despite its small size, it is culturally and linguistically diverse, with people from different ethnic and linguistic backgrounds including Indigenous, African, East Indian, Javanese, Chinese, and European descent.

Suriname is one of the few countries in the world where Dutch is an official language, reflecting its colonial history. The country was a Dutch colony for over three centuries before gaining independence in 1975.

Suriname is home to the largest protected rainforest in the world outside the Amazon Basin. The Central Suriname Nature Reserve covers an area of over 16,000 square kilometers and is home to a diverse array of flora and fauna, including jaguars, giant otters, and numerous species of birds and primates.

The Maroon communities in Suriname are descendants of enslaved Africans who escaped from plantations and established their own autonomous settlements in the interior of the country. Today, the Maroons make up an important part of Suriname's cultural heritage and play a significant role in the country's economy and politics.

The capital city of Paramaribo is a UNESCO World Heritage Site, known for its well-preserved colonial architecture, including wooden buildings from the 18th and 19th centuries. The city is also famous for its vibrant street life, with markets, street vendors, and cultural events taking place daily.

Uruguay

Tango music originated in Uruguay and Argentina in the late 19th century. It is a musical genre that combines African, European, and indigenous musical influences and is known for its sensual and melancholic rhythm.

With a literacy rate of over 98%, Uruguay is one of the most educated countries in Latin America. The country has a long history of free and compulsory education, and the government places a high priority on education.

Mate is a traditional South American drink made from the dried leaves of the yerba mate plant. Uruguay is the world's leading producer of mate and the drink is an important part of the country's culture and social life.

Uruguay is a pioneer in renewable energy and has made significant investments in wind and solar power. The country aims to derive 95% of its electricity from renewable sources by 2030 and is well on its way to achieving this goal.

Uruguay is home to one of the world's oldest democracies, with a history of political stability and peaceful transitions of power. The country has a long history of respect for human rights and is considered to be one of the most socially progressive countries in the world.

Venezuela

Angel Falls, located in the Canaima National Park, is the world's highest waterfall with a height of 979 meters (3,212 feet). The falls are named after Jimmy Angel, a US aviator who was the first to fly over the falls in 1933.

Venezuela has a rich and diverse cultural heritage, with influences from indigenous peoples, African slaves, and European colonizers. The country is known for its vibrant music, dance, and folk art, including the traditional folk music genre known as joropo.

Venezuela is one of the world's largest oil-producing countries and the oil industry is a major contributor to the country's economy. However, the country's oil wealth has been a source of political and economic instability in recent years.

The Parque Nacional de las Tetas de Maria Lionza is a national park in Venezuela that is home to the world's largest dinosaur tracks. The tracks are over 100 million years old and offer a fascinating glimpse into the Earth's distant past.

Venezuela is home to a rich and diverse array of flora and fauna, with many species found nowhere else in the world. The country is a biodiversity hotspot and is home to several national parks and protected areas, including the Gran Sabana and Canaima National Park.

ASIA

Asia is the largest and most populous continent on Earth, covering an area of approximately 44,579,000 square kilometers and home to more than 4.5 billion people. It is located to the east of Europe and north of Africa, and is bordered by the Pacific Ocean to the east, the Arctic Ocean to the north, and the Indian Ocean to the south.

Asia is incredibly diverse, both culturally and geographically. It is home to some of the world's oldest civilizations, including China, India, and Persia, and is also the birthplace of many major world religions, including Buddhism, Hinduism, Islam, and Judaism.

Geographically, Asia encompasses everything from the deserts of the Middle East to the lush tropical rainforests of Southeast Asia, and from the towering peaks of the Himalayas to the vast steppes of Central Asia. There are also many important rivers in Asia, such as the Yangtze, Ganges, Indus, and Yellow Rivers, which have played a crucial role in the development of agriculture and human civilizations in the region.

Asia is also one of the world's fastest-growing economic regions, with countries like China, India, and Japan playing a major role in the global economy. Despite its many challenges, including poverty, conflict, and environmental degradation, Asia remains an incredibly dynamic and influential part of the world, with a rich cultural heritage and a vibrant, diverse population.

There are 55 recognized countries in Asia, however, the number of countries in Asia is a subject of debate and can vary depending on how the boundaries and territories of the continent are defined.

Afghanistan

Afghanistan is a landlocked country located in South Asia and Central Asia. It is bordered by Pakistan to the east and south, Iran to the west, Turkmenistan, Uzbekistan, and Tajikistan to the north, and China to the northeast.

Afghanistan has a rich cultural heritage, with a long history of art, literature, music, and dance. The country is known for its traditional dances, such as the attan, which is performed at festivals and celebrations throughout the year.

Afghanistan is widely considered to be the birthplace of the game of polo, a sport that was played by the Persian nobility as early as the 6th century AD. Today, Afghanistan is home to many polo clubs, and the sport remains an important part of the country's cultural heritage.

Afghanistan has a unique and varied landscape, with a range of mountain ranges, deserts, and fertile valleys. The country is home to the Hindu Kush mountain range, which rises to over 7,000 meters above sea level, as well as the fertile valleys of the Helmand and Arghandab rivers.

Afghanistan has a long and complex history, with a legacy of conflict and political instability stretching back centuries. In recent times, the country has been wracked by conflict, with the United States and its allies invading in 2001 following the September 11 attacks. Despite the ongoing conflict, Afghanistan remains a country of great strategic and cultural significance.

Armenia

Armenia has a rich and diverse cultural heritage, with a history that dates back to over 3,000 years. The country was one of the first to adopt Christianity as a state religion in 301 AD, making it one of the oldest Christian civilizations in the world.

Armenia is a landlocked country located in the South Caucasus region, between the Black Sea and the Caspian Sea. The country is known for its rugged and mountainous terrain, with peaks rising to over 4,000 meters above sea level.

Armenia is famous for its cuisine, with a rich and diverse culinary tradition that is influenced by its history and geography. Some of the country's most famous dishes include khorovats (barbecue), dolma (stuffed grape leaves), and lavash (a type of flatbread).

Armenia has a rich artistic tradition, with a long history of music, dance, and the visual arts. The country is known for its traditional dances, which are performed at festivals and celebrations throughout the year, as well as its intricate and beautiful hand-woven textiles.

Armenia has a rich spiritual heritage, with a long history of religious and spiritual practices. The country is home to many historic churches, monasteries, and other religious sites, which are considered to be some of the most important pilgrimage sites for Christians in the world.

Azerbaijan

Azerbaijan is known as the "Land of Fire" due to its natural fires that have been burning for thousands of years. The Ateshgah Temple, located near Baku, is a testament to this fact and was built by Hindu and Sikh pilgrims who were in awe of the natural flames.

Azerbaijan is the largest country in the Caucasus region and is located at the crossroads of Europe and Asia. It borders the

Caspian Sea to the east and is bordered by Russia to the north, Georgia to the northwest, Armenia to the west, and Iran to the south.

The capital city of Azerbaijan, Baku, is one of the fastest-growing cities in the world and has a rich history that dates back to the 8th century. Baku was once an important center of commerce and culture and was known as the "Paris of the East."

Azerbaijan is famous for its cuisine, which is a blend of Persian, Turkish, and Russian flavors. Some of the most popular dishes include plov (a rice dish with meat and vegetables), dolma (stuffed grape leaves), and bozbash (a soup made with meat and vegetables).

Azerbaijan is also famous for its traditional musical instruments, including the tar (a stringed instrument), kamancheh (a bowed string instrument), and balaban (a type of woodwind instrument). Azerbaijan's traditional music has been recognized by UNESCO as an Intangible Cultural Heritage of Humanity.

Bahrain

Bahrain's archipelago is composed of 33 islands, the largest of which are Bahrain Island, Muharraq Island, and Hawar Island. The country is situated in the Persian Gulf, between Saudi Arabia to the west and Qatar to the east.

The name Bahrain comes from the Arabic word "bahr", which means "sea", and the dual form of the word, "bahrain", means "two seas". This refers to the presence of both freshwater springs and saltwater seas in the country.

Despite its small size, Bahrain has a relatively high population density, with over 1.7 million people living there. The capital and largest city is Manama, which is home to around 200,000 people.

Bahrain is a constitutional monarchy with a king as the head of state and a prime minister as the head of government. The current king is Hamad bin Isa Al Khalifa, who has been in power since 1999.

The pearl diving industry was once the main source of income for Bahrain, with pearls being exported to markets around the world. Today, the country has a modern and diversified economy, with a strong focus on financial services, tourism, and manufacturing.

Bangladesh

The capital and largest city of Bangladesh is Dhaka, which is one of the world's most densely populated cities. Dhaka is a major economic, political, and cultural hub, and is home to many universities, research centers, and government institutions.

Bangladesh is a low-lying country with a predominantly flat terrain, and much of its land area is only a few meters above sea level. As a result, the country is highly susceptible to flooding, which can be devastating for its population and economy. Bangladesh is also vulnerable to cyclones, which can cause extensive damage to infrastructure and property.

Bengali is the official language of Bangladesh and is spoken by the vast majority of the population. English is also widely spoken and used in business, education, and government, and many Bangladeshis are multilingual, speaking multiple languages including Hindi, Urdu, and Arabic.

The majority of Bangladesh's population is Muslim, with over 90% of the population following Islam. However, the country is also home to significant Hindu, Buddhist, and Christian communities, and has a long history of religious and cultural diversity.

Bangladesh is known for its rich cultural heritage, which includes traditional music, dance, and handicrafts. The country is also famous for its cuisine, which features a wide variety of dishes made with spices and herbs, and often incorporates rice and fish as staple ingredients.

Bhutan

Bhutan's unique approach to measuring progress has made it a leader in sustainable development and environmental conservation. The country has adopted policies aimed at preserving its cultural heritage, such as requiring all buildings to be built in traditional Bhutanese style, and limiting the number of tourists who can visit the country each year.

Bhutan is famous for its traditional dress, which includes the gho for men and the kira for women. These garments are made from colorful handwoven fabrics, and are an important part of Bhutanese identity and culture. The gho is a knee-length robe worn by men, while the kira is a long dress worn by women. Both garments are made from woven fabrics, which are often dyed bright colors and decorated with intricate patterns. The patterns and colors of the fabrics are often symbolic, with different designs representing different regions or social classes.

Bhutan is home to a number of rare and endangered species, including the snow leopard, red panda, and black-necked crane. The country has made significant efforts to protect its wildlife and natural environment, and has established a number of national parks and protected areas.

Bhutanese cuisine reflects the country's unique cultural heritage and natural environment. Many dishes feature spicy flavors and a wide variety of herbs and spices, such as ginger, garlic, and cardamom. Rice is a staple of the Bhutanese diet, and is often served with vegetables, meat, or cheese. Ema datshi, a spicy chili and cheese stew, is one of the most popular dishes in Bhutan, and is often served with rice. Another popular dish is momos, a type of steamed dumpling filled with meat or vegetables.

The Bhutanese flag features a dragon, or druk, in the center, symbolizing the country's heritage and its Bhuddist tradition. The flag's background is divided into two parts, with orange representing the spiritual tradition and yellow symbolizing the secular tradition.

Brunei

Brunei is located on the island of Borneo, which is the third-largest island in the world. The island is shared by three countries: Indonesia, Malaysia, and Brunei. The majority of Brunei's population lives in the northern part of the country, which is more urbanized.

Brunei has a well-developed education system, with literacy rates close to 98%. The country provides free education up to the secondary level, and the government also offers scholarships for students to study abroad.

Brunei's economy is heavily dependent on oil and gas exports, which account for over 90% of the country's export earnings. The government has invested heavily in diversifying the economy, particularly in the areas of tourism and services.

The Sultan of Brunei is one of the wealthiest monarchs in the world, with a net worth estimated at over $20 billion. His palace, the Istana Nurul Iman, is one of the largest residential palaces in the world, with over 1,700 rooms.

Brunei is home to a number of unique and endangered animal species, including the Bornean orangutan, the proboscis monkey, and the hornbill. The country has established a number of protected areas to help preserve these species and their habitats.

Cambodia

Cambodia is a relatively poor country, with much of the population relying on subsistence agriculture for their livelihoods. However, the country has seen rapid economic growth in recent years, particularly in the areas of tourism and textiles.

The Khmer Rouge, a communist regime that ruled Cambodia from 1975 to 1979, is responsible for the deaths of an estimated 1.7 million people. The regime was overthrown by Vietnamese forces in 1979, and the country has since struggled to recover from the devastation of the Khmer Rouge era.

Buddhism is the dominant religion in Cambodia, with over 90% of the population practicing the Theravada branch of the faith. The country is known for its elaborate temple complexes, many of which date back to the Khmer Empire.

Cambodia's cuisine is known for its bold flavors and use of fresh herbs and spices. Some popular dishes include amok (a fish curry steamed in banana leaves), lok lak (stir-fried beef with rice), and bai sach chrouk (grilled pork with rice).

Cambodia is home to a number of unique and endangered animal species, including the Irrawaddy dolphin, the Siamese crocodile, and the Asian elephant. The country has established a number of protected areas to help preserve these species and their habitats.

China

China has made significant progress in recent years in terms of environmental protection and sustainability. It is the world's largest producer of renewable energy, including solar and wind power.

Traditional Chinese medicine, including acupuncture and herbal remedies, is an important part of China's healthcare system. Many people in China still rely on traditional medicine for their healthcare needs.

The Chinese New Year, also known as the Spring Festival, is one of the most important holidays in China. It is celebrated with family gatherings, fireworks, and the exchange of red envelopes filled with money.

Chinese art and culture have had a significant influence on the world, including in the areas of painting, calligraphy, ceramics, and literature. China has also made important contributions to science and technology, including the invention of gunpowder, paper, and the compass.

Mandarin Chinese is the most widely spoken language in China, but there are also many other dialects and minority languages spoken throughout the country. Chinese characters are used in writing, and there are over 50,000 characters in the Chinese language.

Georgia

Georgia has a rich history, with evidence of human habitation dating back to the Paleolithic era. It has been ruled by various empires and kingdoms throughout its history, including the ancient Greeks, Romans, and Persians.

Tbilisi, the capital of Georgia, is known for its charming Old Town and mix of architectural styles, including medieval, Art Nouveau, and Soviet-era buildings.

Georgia is home to a number of unique and endangered animal species, including the Caucasian leopard, the brown bear, and the lynx. The country has established a number of protected areas to help preserve these species and their habitats.

The Caucasus Mountains run through Georgia, providing some of the most stunning landscapes in the country. The highest peak is Mount Shkhara, which reaches 5,068 meters.

Georgia is known for its unique and delicious cuisine, which features a variety of meats, vegetables, and spices. Some popular Georgian dishes include khachapuri (cheese-filled bread), khinkali (dumplings filled with meat or cheese), and pkhali (vegetable and nut paste).

India

The Taj Mahal, one of India's most famous landmarks, is made entirely of white marble and was built in the 17th century by the Mughal Emperor Shah Jahan as a mausoleum for his wife, Mumtaz Mahal. The monument is renowned for its stunning beauty and intricate carvings and is a UNESCO World Heritage Site.

India has the world's second-largest population and is expected to overtake China as the most populous country by 2027. The population growth in India is driven by factors such as high fertility rates, improvements in healthcare, and increased life expectancy.

The national animal of India is the Bengal tiger, and the national bird is the peacock. The Bengal tiger is one of the largest and most powerful big cats in the world and is found in the wild in several states across India. The peacock is known for its vibrant colors and striking plumage.

The game of chess originated in India and was originally called chaturanga. The game was played in ancient India with different pieces and rules than the modern game, but the basic principles of strategy and tactics are still the same.

The Himalayan mountain range, which spans over 1,500 miles, is home to some of the world's highest peaks, including Mount Everest. The Himalayas are a vital source of water for India and other countries in the region, and are also home to a rich variety of flora and fauna.

Indonesia

Bali is a popular tourist destination in Indonesia, known for its beautiful beaches, lush rice terraces, and Hindu temples. Other popular destinations include Yogyakarta, which is known for its ancient temples, and the Gili Islands, which are a group of small islands off the coast of Lombok.

The Komodo dragon is the world's largest lizard and is found in Indonesia's Lesser Sunda Islands. The Komodo dragon is a carnivore and is known for its powerful jaws and venomous bite.

Indonesia is the world's largest producer of palm oil, which is used in many food products, cosmetics, and biofuels. The production of palm oil has been a controversial issue, as it has led to deforestation and the displacement of indigenous communities.

Jakarta is the capital city of Indonesia and is one of the world's most populous cities, with over 10 million people. The city is a major economic hub and is home to many skyscrapers, shopping malls, and cultural attractions.

The Rafflesia Arnoldii, which is the world's largest flower, can be found in Indonesia's rainforests. The flower has no stems, leaves, or roots, and relies on a host plant for nutrients.

Iran

Iran has a rich cultural heritage dating back over 5,000 years, with many archaeological sites and artifacts that showcase the country's ancient history. Some of the most famous sites include the ruins of Persepolis, which was the ceremonial capital of the Achaemenid Empire, and the Cyrus Cylinder, which is considered one of the world's first human rights declarations.

Iran's 22 UNESCO World Heritage Sites include some of the most impressive examples of Persian architecture, art, and culture. For example, the historic bazaar of Tabriz is one of the oldest and largest covered bazaars in the world, while the Sheikh Safi al-din Khanegah and Shrine Ensemble in Ardabil is a complex of buildings and tombs that reflects the spiritual and artistic traditions of Iran's Safavid dynasty.

Iran produces over 90% of the world's saffron, which is a spice that is derived from the stigma of the saffron crocus flower. Saffron is prized for its distinctive flavor, aroma, and color, and is used in a wide range of dishes, from rice and meat dishes to desserts and beverages.

Nowruz is a major holiday in Iran and other countries in the region, and is celebrated with a range of traditions and rituals, including the preparation of special foods, the exchange of gifts, and the lighting of bonfires. The holiday marks the start of the new year and the arrival of spring, and is seen as a time of renewal and new beginnings.

Isfahan is a city that is known for its stunning examples of Islamic architecture, including the Imam Mosque, which was built in the 17th century and features a distinctive double-shelled dome, and the Naqsh-e Jahan Square, which is surrounded by historic buildings and is a popular gathering place for locals and tourists.

Iraq

Iraq is home to some of the world's most famous archaeological sites, including the ancient city of Babylon, which dates back to the 18th century BCE, and the ruins of Nineveh, which was the capital of the Assyrian Empire.

The Tigris and Euphrates rivers, which flow through Iraq, were the cradle of civilization for ancient Mesopotamia, and are still important sources of water and irrigation for the country today.

Iraq is known for its rich cultural heritage, including traditional crafts such as pottery, weaving, and metalwork, as well as its music and dance traditions, which include the maqam and the dabke.

Iraq is a major producer of oil, and has one of the largest reserves of oil in the world. The country's oil industry has played a significant role in its economy and politics.

Iraq is home to a diverse population of different ethnic and religious groups, including Arabs, Kurds, Turkmen, and Assyrians, and has a rich cultural tapestry that reflects this diversity.

Israel

Israel's diverse landscapes range from the beautiful beaches of Tel Aviv and Eilat to the snow-capped peaks of Mount Hermon and the stark beauty of the Negev Desert.

Israel is home to several UNESCO World Heritage Sites, including the ancient fortress of Masada, the biblical city of Bethlehem, and the ancient port of Acre.

Hebrew is the official language of Israel, and is the most commonly spoken language. English is also widely spoken, and many Israelis also speak Arabic, Russian, or other languages.

Israel is known for its advanced technology and innovation, with a thriving startup culture that has produced a wide range of groundbreaking products and services in fields such as cybersecurity, mobile apps, and biotechnology.

Israel is a melting pot of cultures and religions, with Jews, Arabs, Christians, Muslims, and Druze all living side by side in a complex and diverse society. While tensions between different groups can sometimes run high, Israelis also have a strong sense of community and mutual support, and the country has a thriving civil society sector that is committed to promoting social justice and equality.

Japan

"Kawaii" culture: Japan is famous for its "kawaii" (cute) culture, which can be seen in the popularity of anime, manga, and cute characters like Hello Kitty.

Japan's traditional arts have been recognized as important cultural treasures, and include practices such as ikebana (the art of flower arrangement), ukiyo-e (woodblock prints), and kendo (a martial art using bamboo swords).

Japanese cuisine is highly varied, reflecting the country's geography, history, and culture. Some other popular dishes include okonomiyaki (a savory pancake), yakitori (grilled chicken skewers), and katsu curry (a Japanese-style curry dish).

Japan is renowned for its technological innovation, and has made significant contributions in fields such as robotics, electronics, and high-speed trains. The country is also a leader in the development of environmentally-friendly technology, including electric vehicles and renewable energy sources.

The Japanese writing system is unique, and includes a mix of kanji (Chinese characters), hiragana, and katakana. Kanji are used for most nouns and adjectives, while hiragana and katakana are used for particles, verb conjugations, and loanwords.

Jordan

Jordan is located in the heart of the Middle East and has a strategic location between Asia, Europe, and Africa. It is also home to several important trade routes, including the ancient Silk Road.

Jordan is a constitutional monarchy, with King Abdullah II serving as the current monarch. The King has significant powers, but also operates within a parliamentary system that includes a Prime Minister and a two-house parliament.

Jordan is home to several famous archaeological sites, including the ancient city of Petra, which is a UNESCO World Heritage Site and one of the New Seven Wonders of the World. The city was carved into the sandstone cliffs over 2,000 years ago, and is renowned for its stunning architecture and intricate carvings.

Jordan's economy is driven by a mix of industries, including tourism, agriculture, and manufacturing. The country is also a significant exporter of phosphate, which is a key ingredient in fertilizer.

Jordan is home to several important religious sites, including the Baptism Site of Jesus Christ on the Jordan River, and Mount Nebo, where Moses is said to have seen the Promised Land. The country is also home to several important Islamic sites, including the ancient city of Madaba, which is home to some of the oldest surviving mosaics in the world.

Kazakhstan

The capital and largest city of Kazakhstan is Nur-Sultan, which was formerly known as Astana. The city is home to several impressive modern architectural landmarks, including the Bayterek Tower and the Astana Opera House.

Kazakhstan is a country of great natural beauty, with a diverse landscape that includes snow-capped mountains, rolling steppe, and vast deserts. The country is also home to several stunning national parks, including the Altyn-Emel National Park and the Kolsai Lakes National Park.

The official language of Kazakhstan is Kazakh, but Russian is also widely spoken in many parts of the country. English and other languages are also spoken by some Kazakhs.

Kazakhstan is one of the world's largest producers of oil and natural gas, and the energy sector is a major contributor to the country's economy. The country is also a significant exporter of metals, including copper and iron ore.

Kuwait

Kuwait is a major player in the global oil industry, with some of the largest oil reserves in the world. The country is a founding member of OPEC (the Organization of the Petroleum Exporting Countries) and has been a key supplier of oil to the world market for many years.

Despite its wealth, Kuwait is also known for its commitment to charitable giving. The country has a long tradition of supporting those in need, and many Kuwaitis donate a portion of their income to charity each year.

Kuwait has a hot and dry desert climate, with temperatures that can reach up to 50 degrees Celsius (122 degrees Fahrenheit) in the summer. Despite this, many Kuwaitis enjoy outdoor activities such as camel racing and falconry.

The official language of Kuwait is Arabic, and Islam is the predominant religion. Kuwait has a rich cultural heritage, with a history that dates back thousands of years.

Kuwait City is the capital and largest city of Kuwait. It is a modern metropolis that is home to many impressive skyscrapers, including the Kuwait Towers and the Al Hamra Tower.

Kyrgyzstan

Kyrgyzstan is a relatively young country, having gained independence from the Soviet Union in 1991. Since then, the country has made great strides in developing its economy and promoting its unique culture and heritage.

Kyrgyzstan is also known for its love of sports, particularly wrestling and horseback riding. The traditional sport of kok-boru, which involves two teams on horseback competing to score goals with a goat carcass, is a favorite among many Kyrgyz.

Bishkek is the capital and largest city of Kyrgyzstan. It is a vibrant city that is home to several impressive cultural and historical landmarks, including the Ala-Too Square, the National Museum of Kyrgyzstan, and the Osh Bazaar.

Kyrgyzstan is home to several stunning national parks, including the Ala Archa National Park and the Sary-Chelek Biosphere Reserve. These parks offer visitors a chance to explore Kyrgyzstan's incredible natural beauty and wildlife.

The country is known for its nomadic heritage, and many Kyrgyz still practice traditional nomadic lifestyles. Visitors can experience this way of life by staying in yurts, a type of portable dwelling that is commonly used by nomads.

Laos

The communist government in Laos came to power in 1975 after a long period of civil war, and has maintained a strict one-party system ever since. However, in recent years there have been some economic and political reforms, including increased openness to foreign investment.

Beer Lao is a popular brand of beer in Laos, and is known for its refreshing taste and low price. It is made from locally-grown rice and is often served ice-cold to combat the hot and humid climate.

The Plain of Jars is a fascinating archaeological site that has puzzled experts for decades. The giant stone jars, which weigh up to several tons each, are believed to be over 2,000 years old and were likely used for burial rituals or as containers for storing rice wine.

The Lao New Year is a festive time in the country, with people pouring water on each other to symbolize purification and renewal. It is also a time for families to gather, exchange gifts, and pay respects to their ancestors.

The Vietnam War had a devastating impact on Laos, with more than two million tons of bombs dropped on the country over a nine-year period. Many of these bombs did not explode and remain a deadly hazard to this day, particularly in rural areas where they can be easily triggered by farmers or children.

Lebanon

The population of Lebanon is estimated to be around 7 million people, and it is one of the most religiously diverse countries in the world. The largest religious groups are Muslims and Christians, with smaller communities of Druze and other religions.

Lebanon has a rich history that spans thousands of years, and the country is home to a number of important historical and archaeological sites. Some of the most notable sites include the ruins of the ancient city of Byblos, the Temple of Jupiter in Baalbek, and the Crusader castle of Tripoli.

Lebanon has a Mediterranean climate with mild, rainy winters and hot, dry summers. The country is known for its beautiful beaches and ski resorts, which are popular with tourists and locals alike.

Lebanon has a diverse and thriving arts scene, with many talented artists and designers based in the country. Beirut, in particular, is known for its street art and graffiti, and the city is home to several galleries and art museums.

Despite its small size, Lebanon has played a significant role in the politics and history of the Middle East. The country has been involved in numerous conflicts and wars, and its geopolitical position has made it a key player in regional affairs.

Malaysia

Malaysia is a constitutional monarchy with a parliamentary system of government. The country is divided into 13 states and three federal territories, with Kuala Lumpur being the capital.

Malaysia is one of the world's leading producers of palm oil, rubber, and tin. The country is also known for its electronics and automotive industries, with many international companies having a presence in Malaysia.

The country has a rich history that is influenced by various cultures, including Indian, Chinese, and European. Malaysia was part of the Srivijayan and Majapahit empires, and later became a British colony before gaining independence in 1957.

Malaysia is known for its delicious and diverse cuisine, which includes Malay, Chinese, and Indian dishes. Some of the most popular Malaysian dishes include nasi lemak, satay, roti canai, and laksa.

Malaysia has a number of popular tourist destinations, including the Petronas Twin Towers in Kuala Lumpur, the Batu Caves, and the tropical islands of Langkawi and Tioman. The country is also home to several national parks and wildlife reserves, including Taman Negara and Bako National Park.

Maldives

The Maldives has a rich history that dates back more than 2,500 years. The country was ruled by various empires over the centuries, including the Buddhist Kingdom of Tamradvipa and the Muslim Sultanate of the Maldives.

The Maldives is home to a number of unique and endangered species, including the Maldives Flying Fox, the Maldives Whistling Thrush, and the Maldives Spiny Lobster.

The Maldives is the only country in the world that has a 100% Muslim population. It is also one of the few countries in the world that is completely carbon-neutral, generating all of its electricity from renewable sources.

The Maldives is known for its traditional music and dance, which is performed during festivals and celebrations. Bodu Beru is a popular form of drumming and singing that is often performed during weddings and other special occasions.

The Maldives is also known for its unique architecture, which features coral stone walls, thatched roofs, and open-air spaces. Many of the traditional Maldivian homes, known as "bodu huts," have been converted into guesthouses for tourists.

Mongolia

The traditional Mongolian diet includes meat, dairy products, and bread. Some of the most popular dishes in Mongolia include buuz (steamed dumplings), khuushuur (fried meat pies), and tsuivan (stir-fried noodles with vegetables and meat).

The Naadam Festival is a major cultural event in Mongolia, featuring competitions in horse racing, wrestling, and archery. The festival is held every summer and attracts thousands of participants and spectators.

Mongolia has a rich literary and artistic tradition, with ancient epic poems, folk songs, and intricate woodcarvings. The country is also home to several modern and contemporary artists, writers, and musicians.

Buddhism is the dominant religion in Mongolia, with around 50% of the population practicing the religion. Shamanism is also a popular spiritual practice, particularly among rural communities.

Mongolia is a democratic country with a parliamentary system of government. The country has made significant progress in recent years, with strong economic growth and improvements in areas such as health and education.

Myanmar (Burma)

The traditional clothing of Myanmar includes the longyi (a type of sarong) and the htamein (a wraparound skirt). Both men and women wear longyis, while the htamein is worn exclusively by women. These garments are made from colorful fabrics and are an important part of Myanmar's cultural heritage.

Myanmar was under military rule for over 50 years, from 1962 until 2011, when the military junta was dissolved and a new civilian government was installed. The country has since undergone significant political and economic reforms.

Myanmar is rich in natural resources, including gems, oil, and natural gas. However, the country's economy has struggled in recent years, with high poverty rates and limited foreign investment.

The official language of Myanmar is Burmese, which is spoken by the majority of the population. There are also several other ethnic languages spoken throughout the country, including Karen, Shan, and Kachin.

Myanmar has a rich cultural heritage, with influences from Buddhism, Hinduism, and Animism. The country is home to a number of important archaeological sites, including the ancient city of Bagan and the Shwedagon Pagoda in Yangon.

Nepal

Nepal is a landlocked country located in South Asia, bordered by India to the south and China to the north. It is home to some of the world's highest peaks, including Mount Everest, which is the highest point on Earth.

Nepal has a diverse landscape, including the Himalayan mountain range, lush forests, and fertile valleys. The country is also home to a number of important rivers, including the Kosi, Gandaki, and Karnali.

The economy of Nepal is largely based on agriculture, with rice, maize, and wheat being the most important crops. The country is also known for its textiles, handicrafts, and tourism industry.

Nepal is a diverse country, with a number of different ethnic groups living within its borders. The largest ethnic group is the Chhetri, followed by the Brahmin, Magar, Tharu, and Tamang.

Nepal has a complex political history, with periods of monarchy, military rule, and democratic government. The country is currently a federal parliamentary republic, with a president as the head of state and a prime minister as the head of government.

North Korea (Democratic People's Republic of Korea)

North Korea has a number of strict rules and regulations, including restrictions on travel, communication, and freedom of expression. The government also enforces strict controls on dress, hairstyles, and other forms of personal expression.

North Korea is known for its large-scale propaganda efforts, which include murals, billboards, and monuments celebrating the country's leaders and ideology.

The country has a number of important cultural sites, including the Kumsusan Palace of the Sun (the mausoleum of Kim Il Sung and Kim Jong Il), the Juche Tower, and the Arch of Triumph.

Despite its isolation, North Korea has been involved in a number of high-profile international incidents, including the sinking of a South Korean naval ship in 2010 and the hacking of Sony Pictures in 2014. The country has also been subject to numerous international sanctions due to its nuclear weapons program and human rights abuses.

North Korea is one of the most secretive and isolated countries in the world, with very limited access for foreigners. The government heavily controls the media and internet access, and foreign journalists are not allowed to operate freely in the country.

Oman

Oman is a country located on the southeastern coast of the Arabian Peninsula, bordered by the United Arab Emirates to the northwest, Saudi Arabia to the west, and Yemen to the southwest. It has a long coastline along the Arabian Sea and the Gulf of Oman.

The economy of Oman is largely based on oil and gas production, with the country being a significant exporter of these resources. The government has also made efforts to diversify the economy, with a focus on tourism and other industries.

Oman is known for its beautiful landscapes, including the Al Hajar Mountains, the Wahiba Sands desert, and the stunning coastline along the Arabian Sea. The country has a number of protected areas, including the Arabian Oryx Sanctuary and the Ras al-Jinz Turtle Reserve.

The capital city of Oman is Muscat, which is known for its beautiful architecture, including the Sultan Qaboos Grand Mosque and the Al Alam Palace. The city also has a number of important cultural institutions, including the Royal Opera House Muscat and the National Museum of Oman.

Oman has a long and rich history, with a number of important historic figures and events. The country was once a powerful seafaring nation, with Omani sailors traveling as far as East Africa and Southeast Asia. The country also played an important role in the early Islamic world, with a number of important Islamic scholars and leaders hailing from Oman.

Pakistan

The official language of Pakistan is Urdu, but there are a number of other languages spoken in the country, including Punjabi, Sindhi, and Pashto. English is also widely spoken and used in business and government.

Pakistan is known for its beautiful landscapes, including the majestic Himalayan Mountains in the north, the rugged Balochistan Plateau in the west, and the fertile Indus River valley. The country is also home to a number of important natural reserves, including the Chitral Gol National Park and the Kirthar National Park.

The capital city of Pakistan is Islamabad, which is known for its beautiful architecture, including the Faisal Mosque and the Pakistan Monument. The city is also home to a number of important cultural institutions, including the Lok Virsa Museum and the National Art Gallery.

Pakistan is a nuclear-armed country and has a large military, with over 650,000 active personnel. The country has been involved in a number of conflicts, including the ongoing conflict in Afghanistan and a number of conflicts with India over the disputed region of Kashmir.

Pakistan is home to a number of important Islamic holy sites, including the Badshahi Mosque and the Data Darbar shrine in Lahore, and the Mausoleum of Muhammad Ali Jinnah in Karachi. The country is also home to a number of important Buddhist and Hindu sites, including the ancient ruins of Taxila and the Katas Raj Temples.

Palestine

Palestine has been the focus of ongoing conflict between Israelis and Palestinians for decades, with both sides claiming the right to self-determination and sovereignty over the land. The conflict has resulted in violence and bloodshed, and has led to the displacement of many Palestinians from their homes.

The economy of Palestine is largely based on agriculture and tourism, but has been negatively impacted by the ongoing conflict and the Israeli occupation of the West Bank and Gaza Strip.

The city of Jerusalem is an important religious and cultural center for both Israelis and Palestinians, and is considered one of the holiest cities in the world. The Old City of Jerusalem is home to a number of important religious sites, including the Western Wall, the Dome of the Rock, and the Church of the Holy Sepulchre.

The Palestinian Authority was established in 1994 as part of the Oslo Accords, and is responsible for governing parts of the West Bank and the Gaza Strip. The authority has limited power, however, due to the ongoing conflict and the Israeli occupation of the territory.

The international community has been involved in efforts to resolve the Israeli-Palestinian conflict, and has called for a two-state solution that would create a Palestinian state alongside Israel. Despite numerous peace negotiations, however, a lasting resolution to the conflict has yet to be achieved.

Philippines

The Philippines was colonized by Spain for over 300 years, and this period had a significant impact on the country's culture and religion. Today, the Philippines is known for its vibrant Catholic traditions, including the observance of Holy Week, which is marked by processions, reenactments of the crucifixion, and other religious ceremonies. Many Filipinos also participate in the "Simbang Gabi" or Misa de Gallo, a series of nine dawn masses that lead up to Christmas Day.

The Philippines is prone to natural disasters such as typhoons, earthquakes, and volcanic eruptions. In 1991, the eruption of Mount Pinatubo was one of the largest volcanic eruptions of the 20th century.

The Philippines has a rapidly growing economy, and is considered one of the "tiger economies" of Southeast Asia. The country's outsourcing industry is one of the largest in the world, with call centers and back-office operations serving clients from around the globe.

The Philippines has a rich history of political activism, and was the site of the People Power Revolution in 1986, which overthrew the authoritarian regime of Ferdinand Marcos.

The Philippines is a tropical country and is blessed with an abundance of beautiful beaches. Boracay, in particular, is one of the most famous tourist destinations in the country. It boasts crystal-clear waters, white sand beaches, and a vibrant nightlife scene. Palawan, on the other hand, is known for its stunning lagoons, limestone cliffs, and coral reefs. Siargao, a popular surfing destination, is known for its big waves and laid-back island vibe.

Qatar

Qatar is a conservative Muslim country, with strict laws governing public behavior and dress. Alcohol consumption is legal but highly regulated, and the sale of pork is prohibited. Women are expected to dress modestly in public, and public displays of affection are frowned upon.

Qatar is home to the world's largest single-site gas-to-liquids (GTL) plant, which is operated by the energy company Sasol. The plant converts natural gas into high-quality diesel and jet fuels using a process called Fischer-Tropsch synthesis. This technology is highly efficient and produces fuels that are cleaner-burning than traditional petroleum-based fuels.

Qatar has a high Human Development Index (HDI) and is considered a developed country by many measures, despite its relatively small size. The country has made significant investments in education, healthcare, and social welfare, which have helped to improve the standard of living for its citizens.

The country has made significant investments in renewable energy and is working to reduce its reliance on fossil fuels in the coming years. Qatar has one of the largest solar energy plants in the world, the Qatar Solar Technologies (QSTec) plant, which produces high-quality solar panels for export.

Qatar is known for its luxurious shopping malls, which feature a range of high-end designer stores and entertainment facilities. The Villaggio Mall, located in the capital city of Doha, is one of the most famous malls in the country. It features an indoor canal with gondolas, a large indoor theme park, and a range of upscale dining options.

Saudi Arabia

The country is home to the two holiest sites in Islam, Mecca and Medina. These cities are visited by millions of Muslims every year during the annual Hajj pilgrimage, which is one of the five pillars of Islam.

The Saudi Arabian economy is heavily dependent on the oil industry, which accounts for around 50% of the country's GDP and 85% of its export earnings. Saudi Arabia has some of the largest oil reserves in the world, and is a major exporter of crude oil.

Saudi Arabia is an absolute monarchy, with the king serving as both the head of state and the head of government. The current king, Salman bin Abdulaziz Al Saud, has been in power since 2015.

The official language of Saudi Arabia is Arabic, and the country follows the Islamic legal system. The government places strict restrictions on free speech and political dissent, and there is little room for opposition or civil society activism.

Women in Saudi Arabia have traditionally had limited rights and opportunities, but in recent years, the government has taken steps to increase gender equality. For example, women are now allowed to drive and are permitted to work in a wider range of professions.

Singapore

Singapore is known for its efficient and well-developed transport system, which includes a modern subway system, buses, and taxis. The country also has a well-connected airport, Changi Airport, which is consistently ranked as one of the world's best airports.

The country has a diverse population, with Chinese, Malay, and Indian ethnic groups making up the majority of the population. English is widely spoken and is the language of business, education, and government.

Singapore is a highly developed country with a strong economy that is heavily focused on manufacturing, trade, and finance. The country has a high per capita income and is often considered a model of economic development.

The country has a well-respected education system, with a strong emphasis on science, technology, engineering, and mathematics (STEM) education. The government provides free primary and secondary education to all citizens, and there are several top universities in the country.

Singapore is known for its strict laws and regulations, which are designed to promote social order and cleanliness. The country has strict laws against littering, spitting, and chewing gum, and there are heavy fines for breaking these laws.

South Korea (Republic of Korea)

South Korea has a highly developed economy, with a strong focus on manufacturing and technology. The country is home to several major companies in the tech industry, such as Samsung and LG.

South Korea is known for its pop culture, especially in the form of K-pop music and Korean dramas. These forms of entertainment have gained a large following both domestically and internationally.

The country has a highly developed transportation infrastructure, including an extensive subway system in Seoul and high-speed rail connections to other parts of the country.

K-pop is a genre of popular music that originated in South Korea and has gained a massive following both domestically and internationally. K-pop music and Korean dramas have become a significant part of South Korea's pop culture and entertainment industry.

The country has a rich history and cultural heritage, including ancient temples, palaces, and fortresses. The city of Gyeongju is often referred to as a "museum without walls" due to its many historical sites.

Sri Lanka

Sri Lanka's name has changed several times over the centuries. It was called Taprobane by the ancient Greeks, Serendib by Arab traders, and Ceylon by the British, who colonized the country in the 19th century. It officially changed its name to the Democratic Socialist Republic of Sri Lanka in 1972, after gaining independence from British rule.

The Sri Maha Bodhi, located in the city of Anuradhapura, is believed to be a sapling of the Bodhi tree under which the Buddha attained enlightenment. The tree is considered one of the most sacred Buddhist sites in the world, and is visited by thousands of pilgrims every year.

Ceylon tea is known for its unique flavor and aroma, which is a result of the island's climate and soil. Sri Lanka's tea industry is one of the country's largest employers, and accounts for a significant portion of its export revenue.

Sri Lanka's UNESCO World Heritage Sites showcase the country's rich history and cultural heritage. The ancient city of Polonnaruwa was the capital of Sri Lanka from the 11th to the 13th century, and is home to impressive ruins such as the Royal Palace and the Gal Vihara, a group of four Buddha statues carved out of granite.

Sri Lanka's national parks and wildlife reserves are home to a variety of animals, including elephants, leopards, and several species of primates. The Sri Lankan elephant is considered a cultural icon and is depicted on the country's coat of arms.

Syria

Syria is home to some of the world's oldest continuously inhabited cities, with Damascus being the oldest. Damascus dates back over 11,000 years and is believed to be one of the oldest continuously inhabited cities in the world. Aleppo and Homs are also ancient cities in Syria that have been inhabited for thousands of years. These cities are known for their rich history, stunning architecture, and vibrant culture.

The city of Aleppo is famous for its souks, which are vast and intricate marketplaces that are made up of a labyrinth of narrow alleys and covered streets. The souks are home to an array of shops selling everything from spices and textiles to gold and silver jewelry. The souks of Aleppo have been in operation for centuries and are considered to be one of the most important cultural and historical sites in the city.

Aramaic is an ancient language that was spoken in the Middle East for thousands of years. Today, it is still used by some Christian communities in Syria, including the Syriac Orthodox Church. The language is an important part of Syrian cultural heritage and is believed to be the language spoken by Jesus Christ and his disciples.

Baklava is a sweet pastry that is popular throughout the Middle East and the Mediterranean. While its origin is not entirely clear, many believe that baklava was invented in Syria. Syrian baklava is known for its light and flaky layers of phyllo pastry, which are filled with a mixture of chopped nuts and syrup.

Syria has a rich archaeological heritage and is home to over 10,000 archaeological sites. These include ancient cities, ruins, and temples, such as Palmyra, Apamea, and Dura-Europos. Palmyra is perhaps the most famous of these sites and is a UNESCO World Heritage Site. The city was once an important trading center and is known for its well-preserved Roman ruins.

Taiwan

Taiwan is known for its street food, and one popular snack is the "stinky tofu," which is a fermented tofu with a strong, pungent odor. Despite the smell, it's actually quite delicious and a popular food item in Taiwan.

In Taiwan, there is a tradition of celebrating the "Ghost Festival," where people make offerings to the spirits of their ancestors. One popular offering is a cake made out of glutinous rice, shaped like a ghost and called "yellow spirit cake."

Taiwan is famous for its night markets, and one of the most popular is the Shilin Night Market in Taipei. Visitors can sample a variety of street food, shop for souvenirs, and even play carnival games.

Taiwan is home to a unique and colorful traditional festival called the "Pingxi Sky Lantern Festival," where people write their wishes on colorful lanterns and release them into the sky.

Taiwan is known for its love of bubble tea, and the country is home to many bubble tea shops where visitors can try this delicious, sweet, and chewy drink. There are also a variety of fun and unique flavors to choose from, including taro, green tea, and mango.

Tajikistan

Tajikistan is home to the famous Pamir Mountains, which are known for their stunning beauty and are a popular destination for adventure seekers.

The country is famous for its traditional music, which is played on instruments such as the dutar and the rubab. The music is an important part of Tajik culture and is often performed at festivals and celebrations.

The traditional Tajik dish of "plov" is a staple food in the country and is made from rice, carrots, and beef or lamb. It is often cooked in a large cast-iron cauldron and served at special events and festivals.

Tajikistan is home to the Fann Mountains, which are known for their unique rock formations and are a popular destination for rock climbing and hiking.

The country is also home to the Zerafshan River, which is a major source of irrigation for the surrounding farmland and is also used for rafting and kayaking by adventure-seekers. The river is surrounded by stunning mountain scenery and is a beautiful spot to relax and take in the natural beauty of Tajikistan.

Thailand

The famous Thai dish "Tom Yum" is a spicy and sour soup that is popular all over the world. The soup is made with a variety of ingredients, including lemongrass, chili, shrimp, and lime juice.

Thailand is home to the Wat Phra Kaew temple, which is considered one of the most sacred Buddhist temples in the country. The temple is known for its beautiful architecture and intricate carvings and is a popular tourist destination.

The famous "Wat Arun" temple in Bangkok is also known as the "Temple of the Dawn" and is one of the most iconic landmarks in the city. The temple is decorated with colorful glass and Chinese porcelain, making it a truly unique and beautiful sight.

Thai massage is a popular form of massage that originated in Thailand and is now practiced all over the world. Thai massage involves stretching, pressing, and massaging the muscles, and is considered to be both therapeutic and relaxing.

The Thai New Year, known as "Songkran," is celebrated in April and is a time for families to come together, cleanse themselves of bad luck, and have fun. During Songkran, people splash each other with water and participate in traditional games and activities.

Timor-Leste (East Timor)

Timor-Leste is the first new sovereign state of the 21st century, having gained independence from Indonesia in 2002.

The country is home to a unique blend of Portuguese and Indonesian culture, as it was a colony of Portugal for over 400 years before becoming part of Indonesia.

Timor-Leste is the only country in Southeast Asia with a majority Roman Catholic population. The Catholic Church plays an important role in the country's cultural and social life.

Timor-Leste is known for its beautiful beaches, including the popular resort town of Atauro Island. The island is a popular destination for snorkeling and diving, and is home to a rich diversity of marine life.

The traditional culture of Timor-Leste is expressed through various forms of dance and music, including the popular "sikka dance." The dance is performed in honor of the country's ancestors and is an important part of Timorese cultural heritage.

Turkey

Turkey is famous for its delicious cuisine, including dishes like kebabs, baklava, and Turkish delight. The country is also known for its strong coffee culture and is a major producer of Turkish coffee.

The city of Istanbul, located in Turkey, spans two continents: Europe and Asia. The city is known for its rich history and cultural heritage, as well as its stunning architecture, including the famous Hagia Sophia and Blue Mosque.

Turkey is home to some of the world's most famous hot springs, including the Pamukkale hot springs. These hot springs are known for their unique terraced formations, which are made of mineral deposits.

The annual Whirling Dervishes ceremony, held in Turkey, is a traditional Sufi religious event that involves dancers twirling in a trance-like state. The ceremony is considered to be a form of meditation and is a beautiful sight to behold.

The Turkish Hammam, or bathhouse, is a traditional form of relaxation and socialization in Turkey. Hammams typically feature hot steam rooms, massage rooms, and social areas, and are a popular destination for both locals and tourists.

Turkmenistan

Turkmenistan is home to the "Gateway to Hell," a natural gas field that has been burning continuously since 1971. The fire is so large that it can be seen from space.

The country is known for its rich tradition of carpet weaving, with Turkmen carpets being highly valued for their intricate designs and quality workmanship.

Turkmenistan has one of the world's largest sand deserts, the Kara-Kum Desert, which covers over 70% of the country.

Turkmenistan is home to several ancient ruins and archeological sites, including the ancient city of Merv, which was once one of the largest cities in the world.

Turkmenistan is known for its unique nomadic culture and traditions, including the "Akhal-Teke horse," a breed of horse that is native to the country and known for its speed and endurance. The horse is considered a symbol of Turkmenistan's national heritage and is highly valued by the Turkmen people.

United Arab Emirates (UAE)

The UAE is home to the world's tallest building, the Burj Khalifa, which stands at over 828 meters (2,722 feet) tall.

The UAE is one of the world's largest producers of oil and natural gas, and its economy is heavily reliant on these industries.

The UAE is known for its unique and luxurious architecture, including the seven-star Burj Al Arab hotel, which is shaped like a sail and is considered one of the most luxurious hotels in the world.

The UAE has a rich cultural heritage, and its capital city, Abu Dhabi, is home to several world-class museums, including the Louvre Abu Dhabi and the Guggenheim Abu Dhabi.

The UAE is a popular tourist destination, known for its beautiful beaches, world-class shopping, and vibrant nightlife. The country also has a strong commitment to sustainability, with several initiatives aimed at reducing its environmental footprint and promoting renewable energy.

Uzbekistan

Uzbekistan is known for its rich cultural heritage, with several ancient cities and monuments, including the city of Samarkand, which is considered one of the oldest cities in the world and is a UNESCO World Heritage Site.

Uzbekistan is one of the largest cotton producers in the world, and cotton is a major part of its economy.

The country is also known for its delicious cuisine, including dishes such as plov (a rice dish with meat and vegetables) and samsa (a meat-filled pastry).

Uzbekistan is home to several unique landscapes, including the Kyzylkum Desert, the Tian Shan Mountains, and the Aral Sea, which was once one of the largest lakes in the world but has shrunk significantly in recent years.

Uzbekistan has a rich musical tradition, with several unique instruments and styles, including the dutar (a two-stringed lute) and Shashmaqam, a classical style of music that originated in

Bukhara. The country is also known for its traditional dance, the "doira," which is performed with a drum of the same name.

Vietnam

Vietnam is the largest producer of coffee in Asia and the second largest producer of rice in the world, after Thailand.

The country is known for its unique and flavorful cuisine, which incorporates a wide range of ingredients and cooking techniques, including grilling, stir-frying, and simmering.

Vietnam is home to several beautiful natural landscapes, including the Ha Long Bay, a UNESCO World Heritage Site known for its stunning rock formations and turquoise waters.

The country has a rich cultural heritage, with several ancient cities and temples, including the Temple of Literature in Hanoi, which was established in 1070 and is dedicated to Confucius.

Vietnam is also famous for its vibrant and bustling cities, including Ho Chi Minh City (formerly Saigon), which is one of the largest cities in Southeast Asia and is known for its bustling street markets, lively nightlife, and delicious street food.

Yemen

The country is also known for its beautiful architecture, including the historic Old City of Sana'a, which is a UNESCO World Heritage Site known for its well-preserved mud-brick buildings and unique tower-houses.

Yemen is home to several unique and endangered species of flora and fauna, including the Arabian leopard and the frankincense tree, which is used to produce the resin used in perfumes and incense.

The country has a rich cultural heritage, with a long history of music, poetry, and storytelling, as well as several traditional festivals, such as the Al-Moulid Festival, which celebrates the birthday of the Prophet Muhammad.

Yemen is famous for its honey, which is considered some of the finest in the world and is used in a variety of traditional dishes, including a sweet bread called Balaleet.

The traditional dance style of Yemen, known as "Ardah," is a highly choreographed, sword-based dance performed by groups of men. The dance is often performed at weddings, festivals, and other special events and is considered an important part of Yemeni cultural heritage.

OCEANIA

Oceania is a region that encompasses the islands of the Pacific Ocean, stretching from the western coasts of the Americas to the shores of Asia and the Indian Ocean. It is made up of a diverse collection of countries, territories, and island groups, including Australasia (Australia and New Zealand), Melanesia (Fiji, Papua New Guinea, and the Solomon Islands), Micronesia (Guam, Kiribati, the Marshall Islands, the Federated States of Micronesia, Nauru, and Palau), and Polynesia (American Samoa, the Cook Islands, French Polynesia, Niue, Samoa, Tokelau, Tonga, Tuvalu, and Wallis and Futuna).

The region is characterized by a rich cultural heritage, with a long history of indigenous cultures, European colonization, and more recent immigration from Asia and other parts of the world. Today, Oceania is home to a diverse and vibrant population, with a wide range of languages, beliefs, and customs.

Oceania is also known for its stunning natural beauty, including pristine beaches, lush tropical forests, and stunning volcanic landscapes. The region is also home to a rich and diverse array of wildlife, including many species of birds, reptiles, and marine mammals found nowhere else on earth.

Oceania is a unique and fascinating region that offers a rich cultural heritage, breathtaking natural beauty, and a vibrant and diverse population.

The exact number of countries in Oceania can vary depending on how the region is defined and the criteria used for determining what constitutes a separate country. However, by most definitions, there are 14 independent countries in Oceania:

Australia

The national emblem of Australia is the kangaroo and the emu, both of which are native to the country and cannot move backwards, symbolizing the nation's progress and forward movement.

Australia is both a country and a continent, and is the world's largest island. It covers an area of 7.68 million square kilometers (2.97 million square miles), making it the world's sixth largest country.

Australia is home to a diverse array of wildlife, including marsupials like kangaroos and koalas, as well as monotremes like the platypus. Many of these species are found nowhere else in the world, and are a unique and fascinating part of Australia's natural heritage.

Australia is the birthplace of surf culture, with the sport of surfing originating in the country in the early 20th century. Today, Australia is home to some of the world's best surfers and boasts some of the most iconic surf spots in the world.

Australia is one of the world's largest wine producers, with the industry contributing billions of dollars to the country's economy each year. The country is known for its high-quality wines, particularly its Shiraz and Chardonnay varietals.

Fiji

Fiji is known for its stunning natural beauty, with crystal-clear waters, white-sand beaches, and lush, tropical forests. The country is a popular tourist destination and is considered one of the most beautiful places in the world.

Fiji has a rich and diverse cultural heritage, with influences from Melanesian, Polynesian, Indian, and European cultures. The country is known for its traditional dances, music, and festivals, as well as its unique arts and crafts.

Fiji is known for its stunning coral reefs, which are home to a diverse array of marine life. The reefs are a popular destination for scuba diving and snorkeling, and are considered some of the most beautiful and pristine in the world.

Fiji is home to a unique language known as Fijian, which is a member of the Austronesian language family. The language has influences from various other Pacific Island languages, as well as from English, Hindi, and other European languages.

Fiji is a popular destination for eco-tourism, with many opportunities for visitors to explore the country's unique and pristine natural environment. From rainforests and waterfalls to coral reefs and beaches, Fiji has something for everyone who loves nature and the outdoors.

Kiribati

Kiribati is the only country in the world that is located in all four hemispheres (northern, southern, eastern, and western), making it a unique and important place for observing the stars and tracking the movement of the sun.

Kiribati is a small island nation located in the central Pacific Ocean, about halfway between Australia and Hawaii. It consists of 33 atolls and reef islands, many of which are sparsely populated or uninhabited.

Kiribati was previously known as the Gilbert Islands, named after British explorer Thomas Gilbert, who first sighted the islands in

1788. The islands were a British protectorate from 1892 to 1979, when they gained independence.

Kiribati is also one of the first places in the world to see the sunrise each day, due to its location close to the International Date Line.

Kiribati is home to the world's largest marine reserve, the Phoenix Islands Protected Area. The reserve covers over 400,000 square kilometers of ocean and is home to a diverse array of marine life, including many rare and endangered species.

Marshall Islands

The Marshall Islands is one of the few places in the world where you can experience the rare phenomenon of "ghost tides," where the tide recedes far out to sea, revealing the shallow waters and coral reefs.

The Marshallese language has its own unique form of counting, where counting is done using the fingers and toes, rather than numbers.

The Marshall Islands is known for its strong traditions of navigation and seafaring, and its people have a rich history of sailing and voyaging across the Pacific Ocean.

The Marshall Islands is home to some of the world's largest coral atolls, including Bikini Atoll, which is known for its incredible diving and snorkeling opportunities and its role in the US nuclear weapons testing program.

The Marshall Islands is made up of 29 coral atolls and 5 single islands, which are spread out over an area of 750,000 square miles, making it one of the largest island nations in the Pacific.

Micronesia (Federated States of Micronesia)

Yap, one of the four states of the Federated States of Micronesia, is famous for its unique currency called "Rai Stones". These are large circular stones that were used as a form of currency in the past and can still be found throughout the island today.

Chuuk, another of the four states of the Federated States of Micronesia, is known for its WWII wreck diving. There are over 60 shipwrecks from the battle of Chuuk Lagoon that are popular tourist attractions for divers.

Kosrae, another state of the Federated States of Micronesia, is known for its lush rainforests and unique flora and fauna.

Pohnpei, the capital of the Federated States of Micronesia, is home to one of the largest archaeological sites in the Pacific. It is believed that the ancient city of Nan Madol was built over a period of 800 years and was the center of political and ceremonial power for the island of Pohnpei.

In Micronesia, traditional stick charts were used for navigation by the indigenous peoples. These charts were made of bamboo sticks and represented the wave patterns of the ocean and the locations of islands.

Nauru

Nauru is the third smallest country in the world by area and the smallest island nation. It covers an area of just 21 square kilometers (8.1 square miles) and has a population of around 10,000 people.

Nauru was once one of the wealthiest nations in the world, thanks to its abundant deposits of phosphate, which is used as a fertilizer. However, the country's phosphate reserves have been depleted and the country's economy is now heavily dependent on foreign aid and investment.

Nauru has a unique culture, with influences from Micronesian, Polynesian, and Melanesian cultures. The country is known for its traditional music, dance, and festivals, as well as its vibrant art and craft scene.

Nauru has a troubled history, with the country facing a number of political and economic challenges in recent years. The country was occupied by various foreign powers, including Germany, Australia, and Japan, and suffered greatly during World War II.

Nauru is a remote and isolated location, located in the middle of the Pacific Ocean. Despite its small size and remote location, Nauru is considered a unique and fascinating destination for travelers who are interested in exploring the world's smallest island nations.

New Zealand

The Lord of the Rings film trilogy, directed by Peter Jackson, was filmed in New Zealand and has since become one of the country's top tourist attractions.

New Zealand is a country with a strong outdoor culture, and adventure activities like bungee jumping, sky diving, and white-water rafting are popular among both locals and tourists.

New Zealand is home to a large population of sheep, with around 10 sheep for every person. In fact, sheep outnumber people in New Zealand by over 6 to 1!

New Zealand is home to a flightless bird called the Kiwi, which is also the country's national symbol. The Kiwi is a nocturnal bird and is so well-known in New Zealand that the people of the country are often referred to as "Kiwis".

The All Blacks, New Zealand's national rugby team, is one of the most successful rugby teams in the world and has a tradition of performing the "haka", a Maori war dance, before every match.

Palau

Palau is one of the few countries in the world that has a "jellyfish lake", where millions of jellyfish have evolved to be completely harmless to humans.

The Rock Islands of Palau are a group of small, uninhabited islands that are famous for their unique rock formations and crystal-clear waters.

Palau is home to a species of giant clam that can weigh up to 600 pounds and can live for over 100 years.

Palau is known for its "Bai", which are traditional meetinghouses that were used for community gatherings and ceremonies. Some of these structures are over a thousand years old and are considered national treasures.

Palau is home to some of the world's most stunning coral reefs, making it a popular destination for snorkeling and diving.

Papua New Guinea

Papua New Guinea is one of the most culturally diverse countries in the world, with over 800 distinct languages spoken across its many islands.

The Huli Wigmen, a tribe from the highlands of Papua New Guinea, are famous for their unique custom of wearing wigs made from their own hair, which they decorate with feathers and flowers.

Papua New Guinea is home to some of the world's largest butterflies, with some species having wingspans up to a foot wide.

The Kamoru Festival is a traditional festival held in Papua New Guinea that features intricate dances and colorful costumes, as well as a variety of food, music, and games.

Papua New Guinea is known for its unique and diverse cuisine, which includes dishes like "saksak", which is made from mashed sweet potatoes and is often served with roasted meats and vegetables.

Samoa

The Fa'a Samoa, or "Samoan way", is a cultural concept that emphasizes respect, hospitality, and community, and is a central part of life in Samoa.

Samoa is home to the world-famous "Siva Samoa" dance, which is performed by men and women and features intricate hand gestures and fluid movements.

The people of Samoa have a strong sense of community, and extended families often live together in large, multi-generational households.

Samoa is located east of the International Date Line, which means that it is one of the first places in the world to see the sunrise each day.

Samoa is famous for its delicious cuisine, which includes dishes like "palusami", which is made from taro leaves wrapped around a filling of coconut cream and onions.

Solomon Islands

The Solomon Islands is an archipelago consisting of over 900 islands located in the South Pacific Ocean. The country has a diverse geography, with rugged mountainous terrain, dense tropical forests, and pristine beaches.

The Solomon Islands has a rich and diverse cultural heritage, with influences from Melanesian, Polynesian, and Micronesian cultures. The country is known for its traditional music, dance, and festivals, as well as its intricate carvings, masks, and other arts and crafts.

The Solomon Islands was a key battlefield during World War II, with a number of key battles and operations taking place in the country. Today, the country is home to several World War II memorials and landmarks, which are popular tourist attractions.

The Solomon Islands has its own unique language, known as Pijin, which is a form of creole based on English and various Melanesian languages. The language is widely spoken in the country and is one of the country's most distinctive cultural features.

The Solomon Islands is a popular destination for eco-tourism, with many opportunities for visitors to explore the country's unique and pristine natural environment. From rainforests and waterfalls to beaches and coral reefs, the Solomon Islands has something for everyone who loves nature and the outdoors.

Tonga

Tonga is a constitutional monarchy, making it one of the few remaining kingdoms in the world. The country is ruled by King Tupou VI, who is regarded as both the political and spiritual leader of the nation.

Tonga is a Pacific island nation located in the South Pacific Ocean. The country is comprised of 176 islands, with the main island of Tongatapu being the center of government and commerce.

Tonga has a rich cultural heritage, with a long history of traditional arts, crafts, music, and dance. The country is known for its elaborate traditional dances, which are performed at festivals and celebrations throughout the year.

Tonga is a popular destination for whale watching, with many opportunities for visitors to see humpback whales up close. The country's clear waters and protected bays make it an ideal location for whale watching, with the best viewing opportunities typically occurring between July and October.

Tonga has a strong sporting culture, with many residents participating in a variety of sports, including rugby, soccer, and boxing. The country has produced several world-class athletes and has a reputation for being a tough and competitive sporting nation.

Tuvalu

The capital of Tuvalu is Funafuti, which is located on the atoll of the same name. Funafuti is the most populated area in Tuvalu, with a population of around 6,000 people. The atoll is home to the country's only airport and is the main transportation hub for the rest of the country.

Tuvalu is one of the smallest countries in the world in terms of land area, with a total land area of just 26 square kilometers. The country's population is also relatively small, with a total population of around 11,000 people.

The country is also one of the lowest-lying countries in the world, with a maximum elevation of just 4.6 meters above sea level. This makes Tuvalu particularly vulnerable to rising sea levels and the impacts of climate change, which threaten to inundate the country's land and freshwater resources.

Tuvalu is a member of the Commonwealth of Nations and has close ties with other Pacific island nations. The country also has a strong cultural connection to nearby countries such as Samoa, Tonga, and Fiji.

The country's economy is largely based on subsistence farming, fishing, and the sale of fishing licenses to foreign vessels. The government has also recently begun exploring the potential for developing the country's renewable energy resources, such as solar and wind power.

Vanuatu

Vanuatu is known for its active volcano, Mount Yasur, which has been erupting continuously for over 800 years and is one of the world's most accessible volcanoes.

The capital city of Port Vila is known for its unique cultural blend, with a mix of Melanesian, French, and British influences.

Vanuatu is home to a number of unique cultural traditions, including the famous "land diving" ceremony, where men dive from towering wooden towers with only vines tied to their ankles.

The country is also famous for its traditional music and dance, which are performed during cultural celebrations and festivals.

Vanuatu is a popular destination for adventure tourism, with a range of activities available, including hiking, snorkeling, and diving.

Additionally, there are several territories in Oceania that are either part of or associated with other countries, including American Samoa, which is an unincorporated territory of the United States, and French Polynesia, which is an overseas collectivity of France.

ANTARCTICA

Antarctica is a unique continent that is governed by the Antarctic Treaty System, an international treaty signed by 54 countries that sets aside Antarctica as a scientific preserve and bans military activity on the continent. While there are no independent countries in Antarctica, several countries, including the United Kingdom, Chile, and Argentina, have made territorial claims on the continent.

However, these claims are not recognized by the international community and have been suspended by the Antarctic Treaty, which designates the continent as a demilitarized zone for scientific research and international cooperation.

In practice, there are no permanent residents or indigenous populations in Antarctica, and the continent is primarily inhabited by scientific researchers and support staff working at research stations operated by various countries.

There are no countries in Antarctica, only scientific research stations and territories claimed by various countries.

SUMMARY

Each country has its own distinct history, culture, and geography. These countries are diverse in their landscapes, languages, and traditions, making the world a vibrant and fascinating place to explore.

One of the largest and most influential countries in the world is the United States of America. With its diverse culture, advanced technology, and political influence, the United States is a major player on the global stage. The country is also known for its stunning natural landscapes, including the Grand Canyon and the Rocky Mountains, as well as its bustling cities, such as New York City and Los Angeles.

Another major country is China, which is the most populous country in the world. With a rich cultural heritage dating back thousands of years, China is known for its historic landmarks, such as the Great Wall of China and the Forbidden City, as well as its thriving economy and political influence.

Europe is also home to a number of important countries, including France, Germany, and the United Kingdom. These countries are known for their rich cultural heritage, including historic landmarks, museums, and art collections, as well as their thriving economies and political stability.

In South America, countries such as Brazil, Argentina, and Peru are known for their diverse landscapes, including the Amazon rainforest, the Andes mountains, and the beaches of Rio de Janeiro. These countries are also home to unique cultures and traditions, including indigenous music and dance styles.

Africa is a continent full of contrasts, with deserts, savannas, and tropical forests all found within its borders. Countries such as South Africa, Nigeria, and Egypt are known for their rich cultural heritage and history, as well as their political and economic influence.

Australia and New Zealand are two countries located in the South Pacific that are known for their unique landscapes and cultures. With stunning beaches, rugged wilderness, and diverse wildlife, these countries offer a wealth of natural beauty and adventure opportunities.

In addition to these larger countries, there are many smaller countries that are also rich in culture, history, and natural beauty. From the tropical islands of the Caribbean to the rugged landscapes of Scotland, each country in the world has its own unique charm and character, making it a fascinating place to explore and discover.

The world is a diverse and fascinating place, with each country offering their own unique histories, cultures, and landscapes. From bustling cities to rugged wilderness, there is something for everyone, making the world a rich and rewarding place to discover and explore.

END NOTE

It's important to note that the facts provided about each country are not an exhaustive list, and there is much more to learn about each country beyond what is mentioned. These fun facts provide a glimpse into some interesting and unique aspects of each country, but they do not represent a definitive guide to all the information or cultural nuances of each country. Therefore, further research and exploration is always encouraged for a more comprehensive understanding of each country.

We hope this brief overview has provided you with a glimpse into the diversity and beauty of the countries of the world. Thank you for taking the time to read this essay, and we hope it has inspired you to learn more about the world and its many fascinating countries.

If you did find this book interesting please share it with your friends and family. Also look out for future publications from Burgerson Publishing.